Charles Ernest Ledoux

Ice-Making Machines

The theory of the action of the various forms of cold-producing or so-called ice

machines

Charles Ernest Ledoux

Ice-Making Machines
The theory of the action of the various forms of cold-producing or so-called ice machines

ISBN/EAN: 9783337163990

Printed in Europe, USA, Canada, Australia, Japan

Cover: Foto ©berggeist007 / pixelio.de

More available books at **www.hansebooks.com**

ICE-MAKING MACHINES:

THE THEORY OF THE ACTION OF THE VARIOUS FORMS OF COLD-PRODUCING OR SO-CALLED ICE MACHINES
(MACHINES A FROID).

TRANSLATED FROM THE FRENCH OF

M. LEDOUX,
Ingenieur dès Mines.

REPRINTED FROM VAN NOSTRAND'S MAGAZINE.

NEW YORK:
D. VAN NOSTRAND, PUBLISHER,
23 MURRAY AND 27 WARREN STREET.
1879.

PREFACE.

The theory of Ice-Making Machines has assumed a new importance, since it has been shown that they may be worked to an economical advantage in some sections, even where natural ice is not difficult to be obtained.

But aside from any question of competition with natural ice in temperate climates, the subject is of great interest to those who find it desirable to produce and maintain a low temperature in places where the requisite quantity of ice would be too cumbersome, and where a refrigerating machine and its driving power can be easily accommodated. Such an example is afforded by the hold of a vessel sailing in a warm climate.

The conditions of effective working of the three classes of machines are clearly set forth in this little treatise.

G. W. P.

ICE-MAKING MACHINES.

Chapter I.

§ 1. It has long been known that air is heated or cooled when compressed or dilated.

The mechanical theory of heat defines the conditions under which this heating or cooling is effected, and shows that these effects are proportioned to the external work performed by the air, with the restriction that in expanding the resistance overcome by the gas is always equal to the elastic force of the latter.

If t and t' represent successive temperatures of a unit weight of a permanent gas, which has been compressed or dilated under conditions above stated in producing an amount of work (either resistant or motive) equal to W, we shall have

$$t - t' = \frac{A}{c} W$$

A being the reciprocal of the mechanical equivalent of heat $=\frac{1}{424}$ and c being the specific heat of the gas at constant volume.

In a saturated vapor a part of the thermal equivalent of the external work is transformed into latent heat; the other part alone becomes sensible under the form of external heat.

This is expressed in the fundamental equation

$$c_1(t-t') + (x\rho - x'\rho') = AW$$

in which c_1 is the specific heat of the liquid, x the proportion of vapor in the unit of weight of mixture of liquid and vapor, ρ the latent heat of the vapor and W the external work accomplished.

We see from these equations that for the same quantity of heat transformed into work, the range of temperatures must be greater with a gas than with saturated vapors.

§ 2. Whether we employ a permanent gas or a vapor, the apparatus designed for the refrigerating effects is based upon the following series of operations:

Compress the gas or vapor by means of some external force, then relieve it of its heat so as to diminish its volume; next, cause this compressed gas or vapor to expand so as to produce mechanical work and thus lower its temperature. The absorption of heat at this stage by the gas, in resuming its original condition, constitutes the refrigerating effect of the apparatus.

When the cooling takes place at constant pressure, the cycle of operations can be represented by the diagram Fig. 1 in which the abscissas represent volumes, and the ordinates pressures.

The gaseous body taken at the pressure P_0 and under the volume V_0 is compressed to the tension P_1 and the volume V_1. It is then cooled under constant pressure so that the volume V_1 becomes V_1', then it is allowed to expand, the pressure P_1 becoming P_0 and the volume changing from V_1' to V_2. Finally it is brought to the original volume V_0 by transferring heat to it under constant pressure. The area $V_0 V_1 V_1' V_2$ represents

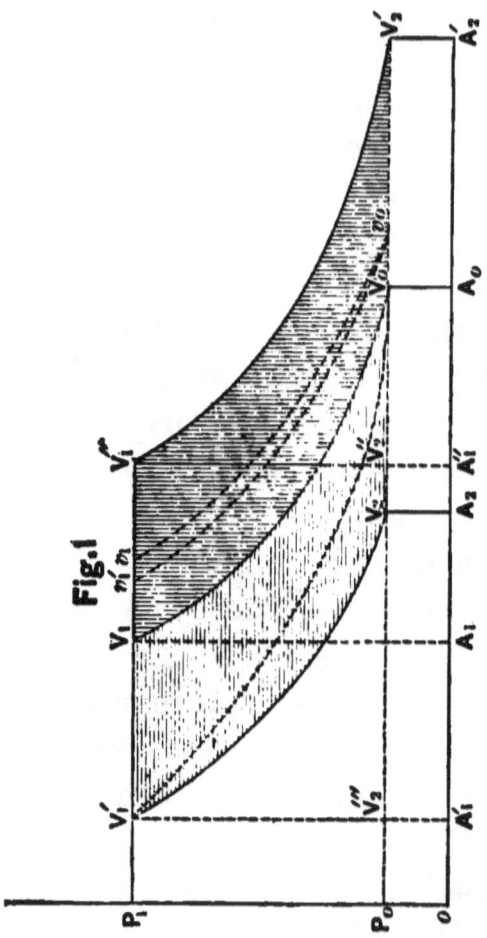

the work expended and the line V_0V_2 the refrigerating effect obtained.

An inspection of the figure shows that a refrigerating machine is a heat engine reversed.

If instead of cooling the gas, to reduce it from the volume V_1 to V_1', it be heated so as to assume the volume V_1'' greater than V_1 an amount of work is obtained which is represented by the vertically shaded area $V_0V_2'V_1''V_1$; the heat expended is represented by the length V_1V_1''.

It should be noticed that in the case of a permanent gas, the changes from volume V' to V_1' or V_1'' and from V_2 or V_2' to V_0 are accompanied by corresponding changes in temperature. In the case of a condensable vapor these changes are effected at a constant temperature, the addition or subtraction of heat taking effect in an evaporation of the liquid or a condensation of the vapor.

§ 3. From this similarity between heat motors and freezing machines it results that all the equations deduced from the mechanical theory of heat to determine the performance of the first apply equally to the second.

If Q_1 be the quantity of heat taken from or added to a given mass, of compressed gas or vapor, and Q the quantity of heat necessary to subtract from or add to the expanded mass in order to bring it to its initial state, T_0 and T_1 the absolute temperatures corresponding to the volumes V_0 and V_1 and W the work, either active or resistant developed by the machine. The fundamental principle of the mechanical theory of heat, if the gas returns exactly to its primitive condition, affords the equation,

$$Q_1 - Q = AW$$

If the cycle of changes is the so-called cycle of Carnot; that is to say, if the lines V_1V_0, $V_1'V_2$, and $V_1''V_2'$ are adiabatic curves; then we have

$$\frac{Q}{T_0} = \frac{Q_1}{T_1} = \frac{Q_1 - Q}{T_1 - T_0}$$

The quantity of work developed by a heat motor, under these circumstances, is for each heat unit or *calorie*, whatever the intermediate agent,

$$\frac{W}{Q_1} = \frac{1}{A} \cdot \frac{T_1 - T_0}{T_1}$$

The efficiency depends upon the difference between the extremes of temperature.

The performance of a refrigerating machine depends upon the ratio between the calories eliminated and the work expended in cooling.

It is expressed by

$$\frac{Q}{W}$$

and we have

$$\frac{Q}{W} = \frac{AQ}{Q_1 - Q} = A \frac{T_0}{T_1 - T_0}.$$

This result is independent of the nature of the body employed.

Unlike the heat motors, the freezing machines possess the greatest efficiency when the range of temperatures is small, and when the final temperature is elevated.

In a freezing machine employing a vapor, T_0 being the absolute minimum final temperature, this final temperature T_2 in a machine employing a permanent gas is different from the initial temperature T_0, and we have,

$$\frac{T_1}{T_0} = \frac{T_0}{T_2}$$

We can write for the efficiency

$$\frac{Q}{W} = A \frac{T_2}{T_0 - T_2}$$

Comparing the efficiencies of the two machines it is evident that the performance becomes less in proportion as we obtain lower final temperatures.

Theoretically there is no advantage in employing a gas rather than a vapor in order to produce cold even if the compression be made without addition or subtraction of heat.

The choice of the intermediate body would be determined by practical considerations based on the physical characteristics of the body, such as the greater or less facility for manipulating it; the extreme pressures required for the best effects, etc.

Air offers the double advantage that it is everywhere obtainable, and that we can vary at will the higher pressures independent of the temperature of the refrigerant. But it is cumbersome, and to produce a given useful effect the apparatus must be of large dimensions.

Liquids on the other hand allow the use of smaller machines, but are obtained only at a greater or less cost.

Furthermore the maximum pressure is determined beforehand by the temperature of the refrigerant, and depending on the nature of the volatile liquid; this pressure is often very high.

§ 4. The foregoing conclusions are based on the hypothesis that the compression and expansion follow the adiabatic lines $V_0 V_1$ and $V_1' V_2$, that is to say that the changes of volume and pressure follow the cycle of Carnot.

This hypothesis is realized when the cooling is accomplished outside of the compression cylinder and after the gas has been raised to the pressure P_1.

If the compression be effected according to some cycle different from Carnot's, the efficiency, if it be a heat motor, would be diminished, but in a freezing machine it would be greater or less, depending upon the manner in which the successive operations were effected.

Suppose for example that instead of cooling, the gaseous body outside the compression cylinder, it be done during compression within the cylinder in such a manner as to maintain a constant temperature. This hypothesis would be graphically represented in Fig. 1 by replacing the adiabatic curve V_0V_1 by the isothermic curve V_0V_1'. The work of resistance of the machine would then be represented by the curvilinear triangle $V_0V_1V_1'V_2$. The quantity of negative heat produced represented by the line V_0V_2 remains the same. The efficiency of the freezing machine would be thus augmented as the resistant work of the motor would be less than the preceding case for the same quantity of negative heat produced.

The cooling of vapors during compression is not readily realized, since it is effected at a constant temperature and one which is lower than the refrigerant. It is realized though somewhat incompletely in the case of permanent gases since their temperature during compression is above that of the refrigerant.

§ 5. The efficiency is calculated in the following manner.

We suppose the compression to be made at a constant temperature. Then by Marriotte's Law we have $P_1 V_1 = P_0 V_0$.

The work of resistance to compression would be

$$W_r = P_0 V_0 . l \frac{V_0}{V_1} = RT_0 l \frac{V_0}{V_1}$$

and we shall have as in the preceding case.

$$AW_r = Q_1$$

R is a constant, uniform for the air at 29.27 inches and a unit of weight is supposed taken.

The gas dilating from the temperature T_0 to T_2 without gaining or losing heat, we shall have for the work of dilatation, inclusive of the work at full pressure during introduction;

$$AW_m = kc(T_0 - T_2) = Q.$$

The performance is represented by

$$A \frac{Q}{Q_1 - Q}$$

and we have

$$\frac{Q}{W_r - W_m} = A \frac{Q}{Q_1 - Q}$$
$$= \frac{kc(T_0 - T_2)}{RT_0 l \frac{P_1}{P_0} - \frac{kc}{A}(T_0 - T_2)}.$$

We have also

$$\frac{c}{A} = \frac{R}{k-1};$$

k is the ratio of specific heat at constant pressure to the specific heat at constant volume; this ratio is $=1.41$ and is the same for all permanent gases.

It follows then

$$A \frac{Q}{Q_1 - Q} = A \frac{T_0 - T_2}{\left(\frac{k-1}{k}\right) T_0 l \frac{P_1}{P_0} - (T_0 - T_2)}$$

If the compression follows an adiabatic curve, we shall have for the efficiency—

calling T_1 the absolute final temperature of the compression

$$A\frac{Q}{Q_1-Q} = A\frac{T_0-T_2}{T_1-T_0-(T_0-T_2)}$$

and $$\frac{T_1}{T_0} = \left(\frac{P_1}{P_0}\right)^{\frac{k-1}{k}}$$

It is easy to show that

$$T_1-T_0 \text{ or } T_0\left\{\left(\frac{P_1}{P_0}\right)^{\frac{k-1}{k}}-1\right\}$$

is greater than

$$\frac{k-1}{k}T_0 l\frac{P_1}{P_0}$$

and consequently that the efficiency in the first case is less than in the second.

The employment of air presents a certain theoretical advantage over volatile liquids, inasmuch as it admits of cooling to a certain extent during compression. We will now examine in succession some of the recently invented freezing machines (*machines a froid*). The Air Machine of M. Giffard; the Sulphurous Acid Machine of M. Pictet, and the Ammonia Machine of M. Carré.

Chapter II.

GIFFARD'S AIR MACHINE.

§ 7. This machine consists of a single-acting cylinder A, the piston of which is furnished with two valves opening from without inward. This cylinder is surrounded with a jacket leaving a space within which circulates a current of cold water.

There is a second cylinder, B, also single-acting, and having a solid piston, and with a diameter a little smaller than the first. At the bottom of this cylinder are two openings closed by valves, opening, one outward and the other inward, and operated by levers which are worked by cams on the driving shaft.

The pistons are driven by crank connections with the main shaft.

The condenser R is a surface condenser and receives a current of cold water from the envelope of the compressor cylinder A. A Reservoir of wrought iron, R', is connected with the condenser by a tube and communicates also with the bottom of the expansion cylinder B.

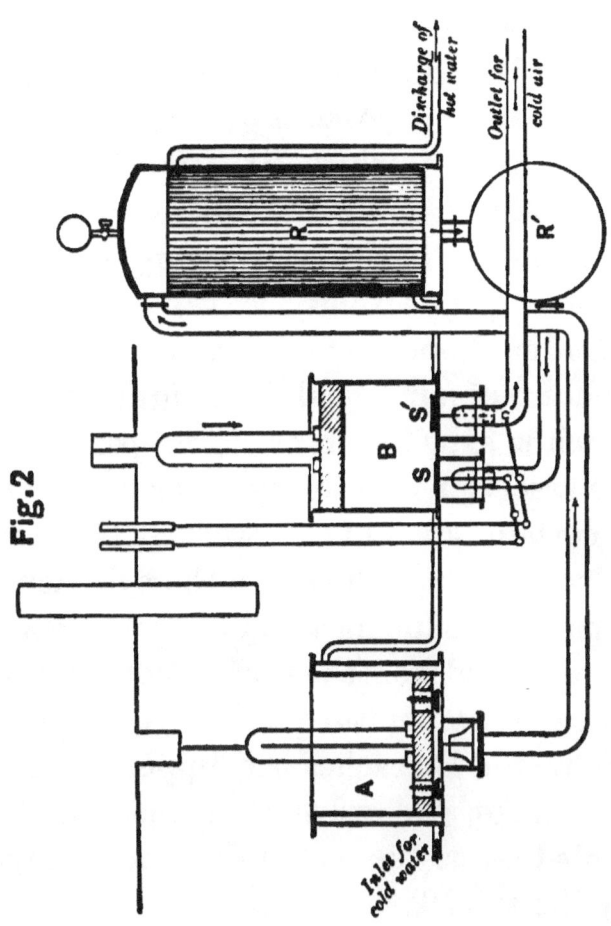

§ 8. The air taken in at ordinary pressure is compressed in the cylinder A till it has the density of that in the reservoir; it is then allowed to flow into the condenser R and the reservoir R'. During this passage it loses a great part of the sensible heat which it attains during compression, and is brought nearly to the temperature of the surrounding air.

During this time the valve s of the cylinder B opens and permits a certain amount of air equal in weight to that which is expelled from A, to pass from the reservoir into the cylinder producing a certain amount of work. Then the valve s closes,—the air in the cylinder B expands producing again work which may be deducted from the work of compression and the temperature is lowered. When the piston B reaches the upper limit of its stroke, the valve s' opens and the cooled air as the piston descends escapes by the tube T.

The cooling experienced by the air, during compression, by contact with the cooled sides of the cylinder is scarcely sensible.

The machine therefore acts under conditions set forth in § 2 and we know that its useful effect cannot exceed the value

$$A\frac{T_0}{T_1-T_0} \quad \text{or} \quad A\frac{T_2}{T_0-T_2}$$

By means of the adjustable cams we can regulate at will the action of the valves s and s'. If we shorten the time of admission into the cylinder B, the pressure will increase in the reservoir; for the amount flowing into B should be equal to that forced into the reservoir from A. The temperature of the air expelled will then be less. If, on the contrary, we increase the time of admission the reservoir pressure will diminish, and the temperature of outflowing air will be increased.

The apparatus presents then this important peculiarity—that we can vary the useful effect of the machine at will, through wide limits.

As the air leaves B, at the pressure of the atmosphere, the minimum limit of pressure is established, below which the expansion cannot be pushed, and which is controlled by the relative dimensions of the two cylinders.

We will proceed to calculate the cooling effect produced by this machine and the corresponding work required. We shall neglect at first the effect of waste spaces in the machine, and of watery vapor in the air.

§ 9. Let P_0, t_0 and T_0 be the pressure and temperature (counted from absolute zero) of the air.

V_0 the volume described by the piston A.

V_1 the volume of air when at pressure P_1.

V_1 is then the volume described by the piston during the outflow.

m = weight of air whose volume passes from V_0 to V_1.

P_1, t_1 and T_1 the pressure and temperature of compressed air delivered from A.

V_1' t_1' and T_1' the volume and

temperature after passing into the condenser.

V_2 the total volume described by piston B.

P_2, t_2 and T_2 the pressure and temperature of the air at the end of the course of this piston.

During compression the cooling by simple contact with the sides of the cylinder is insignificant. We shall neglect this and also assume that no heat is received from the sides of the cylinder B.

FIRST PERIOD: COMPRESSION.

§ 10. When air is compressed without losing or gaining heat, the pressure and temperature at each instant bear the relation to each other expressed by the equation

$$P_0 V_0^k = P_1 V_1^k \qquad (1)$$

in which k is the ratio of specific heat of constant pressure to the specific heat of constant volume.

$$k = \frac{0.23751}{0.16844} = 1.41$$

Gay Lussac's law affords,

$$P_0 V_0 = R m T_0 \qquad (2)$$

and
$$P_1 V_1 = R m T_1 \qquad (3)$$

From equations 1 2 and 3 we deduce

$$\frac{T_1}{T_0} = \left(\frac{P_1}{P_0}\right)^{\frac{k-1}{k}} \qquad (4)$$

$$\frac{T_1}{T_0} = \left(\frac{V_0}{V_1}\right)^{k-1} . \qquad (5)$$

The work of the resistance to compression and outflow is

$$W_r = \frac{k}{k-1}(P_1 V_1 - P_0 V_0). \qquad (6)$$

We have elsewhere

$$\frac{k}{k-1} = \frac{kc}{AR}.$$

c being the specific heat of air of constant volume.

Equation (6) then becomes

$$W_r = \frac{mkc}{A}(T_1 - T_0). \qquad (7)$$

SECOND PERIOD: COOLING.

The air is cooled in the condenser under constant pressure. The volume changes from V_1 to V_1', and the temperature from t_1 to t_1'.

we have;
$$V_1' = V \frac{T_1'}{T} \qquad (8)$$

and the quantity of heat imparted to the water of the condenser is;
$$Q_1 = mkc(T_1 - T_1') \qquad (9)$$
If $T_1' = T_0$ then $R_1 = AW_r$

THIRD PERIOD; EXPANSION.

The volume V_1' of air enters the cylinder B yielding an amount of work equal to P_1V_1'. It expands from V_1' to V_2 without gain or loss of heat. We have then:

$$P_1 V_1'^k = P_2 V_2^k, \qquad (10)$$
$$P_1 V_1' = RmT_1', \qquad (11)$$
$$P_2 V_2 = RmT_2 \qquad (12)$$

whence
$$T_2 = T_1' \left(\frac{P_2}{P_1}\right)^{\frac{k-1}{k}}. \qquad (13)$$

The work performed by the air is

$$W_m = \frac{k}{k-1}(P_1 V_1' - P_2 V_2) \qquad (14)$$

or
$$W_m = \frac{mkc}{A}(T_1' - T_2) \qquad (15)$$

The resistances to be overcome by external force amount to

$$W_r - W_m = \frac{mkc}{A}[(T_1 - T_1') - (T_0 - T_2)]. \quad (16)$$

If the machine works properly, the final pressure P_2 should be equal to the atmospheric pressure.

The equations (10) (12) and (13) give

or
$$\left. \begin{array}{c} \dfrac{V_2}{V_1'} = \dfrac{V_0}{V_1} \\ \dfrac{V_2}{V_0} = \dfrac{T_1'}{T_1} \end{array} \right\} \quad (17)$$

and
$$\frac{T_2}{T_1'} = \frac{T_0}{T_1} \quad (18)$$

Equation (17) expresses the ratio which should exist between the volumes of the two cylinders, in order that the air be finally expelled at atmospheric pressure, after having been compressed by a force P_1.

The negative heat (cooling), produced by the apparatus, is the quantity of heat necessary to restore the air from the temperature t_2 to the temperature t_0, under constant pressure.

or
$$Q = mkc(T_0 - T_2) \\ Q = mkcT_0\left(1 - \frac{T'_1}{T_1}\right)} \quad (19)$$

§ 11. Since a given weight of air is restored, at the end of the operation, to the same temperature and pressure it had at the beginning it follows, that it has been through a perfect cycle and we have from the mechanical theory of heat;

$$Q_1 = A(W_r - W_m) + Q$$

The theoretical performance of the machine is, calling it u,

$$u = \frac{Q}{W_r - W_m} = A \cdot \frac{T_0 - T_2}{(T_1 - T'_1) - (T_0 - T_2)},$$

$$u = A \cdot \frac{1}{\frac{T_1 - T'_1}{T_0 - T_2} - 1}$$

and as we have from equation (18)

$$\frac{T_1 - T'_1}{T_0 - T_2} = \frac{T_1}{T_0} = \frac{T'_1}{T_2}$$

we get finally

$$u = A \cdot \frac{T_0}{T_1 - T_0} = A \cdot \frac{T_2}{T'_1 - T_2}, \quad (20)$$

a result already found in § 3 by supposing $T_1'=T_0$. If $T_1>T_0$ the useful effect is diminished.

The efficiency of the machine will be all the greater as T_1 approaches in value to T_0; that is to say as it is urged at a lower pressure into the reservoir. But as we lower the pressure of working, the quantity of negative heat produced diminishes also and becomes nothing when $T_1'=T_1$.

The necessary driving power W_r-W_m which we proceed to calculate, should be augmented by the passive resistances.

If we consider the refrigerating machine as composed of two distinct machines driven by the same shaft, we are led to consider that the work of the passive resistances is proportional not to the final work W_r-W_m but rather to the sum of the work developed in the two cylinders W_r+W_m. Considering the simplicity of the machine, the small amount of friction, and the absence of a stuffing box, we can admit that the work of the passive resistances should not exceed eight per cent of the above total work.

The resistance of the machine is then $1.08 W_r - 0.92 W_m$.

The following table gives the amount of refrigeration obtained, and the work expended, by passing a cubic meter of dry air through the machine; the pressures in the reservoir varying from $1\frac{1}{2}$ to $4\frac{1}{2}$ atmospheres. The temperature of the external air is taken at $15°$; the temperature of the air leaving the condenser at $18°$; temperature of the water about $13°$ $V_0 = 1$, $T_0 = 288$ and $m = 1^k.266$.

§ 12. An examination of the table shows the enormous influence that the passive resistances exert upon the efficiency of air machines. It is one of the consequences of the inherent cumbrousness which follows from the use of this body in a thermic machine.

The useful effect produced is not increased in proportion to the increase of pressure. It is of no advantage to employ pressures higher than about $4\frac{1}{2}$ atmospheres. Aside from the diminution of efficiency of the air at high pressures, a loss is occasioned by heat developed in

P_1 (Atmospheres.)	Temperature of air after compression. t_1.	Temperature of outflowing air. t_3.	Diminution obtained. t_0-t_3.	Negative calories obtained.	Work expended, theoretical kilogrammeters.	Work expended, effective.	No. of negative calories developed.			
							Per kilogrammeter, theoretic.	Per horse power per hour, theoretical.	Kilogrammeter really expended.	Per horse power per hour.
	degrees	degrees	degrees	cal.			cal.	cal.	cal.	
1½	51,04	—14,36	29,36	8,548	454,26	1.070,5	0,01882	5.081,4	0,007985	2.156
2	79,31	—35,12	50,12	14,591	1.381,26	2.481,6	0,01056	2.851,2	0,00588	1.587
2½	102,93	—49,55	64,55	18,792	2.515,64	3.991,1	0,00747	2.016,9	0,00471	1.281
3	123,39	—62,58	77,58	22,586	3.556,3	5.412,3	0,00635	1.714,8	0,00417	1.126
3½	141,57	—70,84	85,84	24,991	4.657.3	6.724,2	0,00537	1.448,8	0,00372	1.003
4	157,98	—78,54	93,54	27,232	5.732,5	8.097,0	0,00475	1.282,6	0,00336	907
4½	173,00	—85,09	100,09	29,375	6.778,0	9.297,0	0,00433	1.170,2	0,00316	853

the compressor, and which extends to other working parts of the machine. We have said above that, with a given machine we can vary at will the pressure P_1 by varying the length of time of the opening of the admission valve in the cylinder B. If the time be shortened the pressure and the cooling effect are both increased; and if the time be increased P_1 is diminished. It is necessary that we should vary at the same time the working of the emission valve, so that it opens at the moment when the piston shall have passed through a space equal to $V_0 \frac{T_1'}{T_1}$ corresponding to the atmospheric pressure on the inside of the expansion cylinder.

A machine whose dimensions and velocity are such that it uses 1000 cubic meters of air per hour will produce from 8.548 to 29.375 negative calories and upwards per hour, provided that the driving power varies from 4 to 34 horse power.

Practically however the efficiency of air machines is not so great as is indicated by the above table as no account has yet been taken of watery vapor in the air, nor of lost spaces in the machine.

We proceed to examine the influence of these two causes of loss.

INFLUENCE OF MOISTURE IN THE AIR.

§ 13. This influence is not to be neglected. The vapor contained in the air condenses on the sides of the expansion cylinder, and parts with its latent heat of vaporization so that the final temperature of the air is higher than it would have been if dry.

Furthermore the snow produced from this moisture accumulates around the orifice of the cold air outlet and we cannot readily utilize the cold which is required to produce it. For these two reasons, but especially for the latter, the moisture of the air causes a notable loss.

We proceed to calculate the volume and the temperature of the air at the end of the expansion under the supposition of a known hygrometric state of the atmosphere, from which we can easily deduce by the tables the pressure of the vapor p_0 and its weight μ_1.

In the compression cylinder of watery vapour not being near the saturation point, and exerting a feeble pressure will behave nearly as a perfect gas; its volume and its temperature are represented by the relations $pv^k = $ a constant, in which $k=1.41$ and $pv=\text{R}'m\text{T}$;

$$\text{R}' = \frac{\text{R}}{0.622} = 47.061.$$

The total pressure of air and vapor being represented by P, the pressure of the vapor being p, that of the air alone will be $\text{P}-p$ and we shall have preserving our former notation:

$$P_1 V_1^k = P_0 V_0^k, \qquad (21)$$
$$p_1 V_1^k = p_0 V_0^k, \qquad (22)$$
$$(P_0 - p_0)V_0 = Rm T_0, \qquad (23)$$
$$p_0 V_0 = R'\mu_1 T_0, \qquad (24)$$
$$(P_1 - p_1)V_1 = Rm T_1, \qquad (25)$$
$$p_1 V_1 = R'\mu_1 T_1. \qquad (26)$$

$$W_r = \frac{k}{k-1}(P_1V_1 - P_0V_0)$$

or $$W_r = \frac{A}{k}(mc + \mu_1 c')(T_1 - T_0)$$ (27)

c' is the specific heat under constant volume of the superheated vapor

$$c' = 0{,}3407.$$

After cooling the volume becomes

$$V'_1 = V_1 \frac{T'_1}{T_1} \qquad (28)$$

and we have
$$p_1 V'_1 = R' \mu_1 T'_1.$$

From equations 21 and 22 we can deduce the pressure in the reservoir.

We can determine by examining a table of tensions of saturated steam whether the pressure p_1 is greater or less than the pressure which corresponds to the temperature T'_1. If it be less the air will not be saturated with vapor when leaving the condenser, and the heat absorbed by the latter will be:

$$Q_1 = k(mc + \mu_1 c')(T_1 - T'_1)$$

If the pressure p_1 is greater than the

pressure p_1', corresponding to the temperature T_1' for saturated steam, there will be a condensation of some of the vapor in the condenser; the amount condensed will be
$$\mu_1(1-x_1')$$
and the pressure of the vapor entering into the cylinder B will be p_1', that of the air being P_1-p_1'.

We shall have also:
$$x_1' = \frac{p_1'}{p_1} = \frac{p_1'}{p_0} \cdot \frac{P_0}{P_1}$$

We see that the quantity of vapor not condensed by the cooling, and passing into the expansion cylinder, will continually diminish in proportion as the working pressure is raised. The influence of the humidity in the air will therefore be less as the pressure is made greater.

The weight of the mixture of air and vapor, which is $m+\mu_1$ if there is no condensation in the cooler or $m+\mu_1 x_1'$ if there is a condensation, is carried into the cylinder B where it encounters the surfaces cooled during the preceding stroke. We can neglect the influence of these cold surfaces upon the air alone, but not upon the mixture of air and vapor. The latter is converted into frost which releases a certain amount of heat to be imparted to the metal, and which during the expansion is restored to the air.

Suppose at first that there is no condensation in the cooler, there is conveyed to the cylinder a weight μ_1 of saturated, or nearly saturated, vapor at the temperature T_1'. We may assume, considering the very low temperature of the surfaces, that all the vapor is condensed here; it will disengage a quantity of heat C, which is approximately equal to $\mu_1(r_1' + 79)$. r_1' being the latent heat of the vapor corresponding to the temperature t_1', 79 is the latent heat of water released on freezing.

The heat C is gradually restored to the air during expansion.

The pressure of the air becomes P_1, and the volume introduced into the cylinder is

$$V_1'' = \frac{RmT_1'}{P_1}.$$

The differential equation of the work is

$$\frac{c}{A} m\, dT + \frac{c_1}{A} \mu_1 dT - \frac{dC}{A} = -P\, dV$$
$$= -RmT\, \frac{dV}{V};$$

c_1 being the specific heat of ice, $= 0,5$

or $\left(\dfrac{c}{AR} m + \dfrac{c_1}{AR} \mu_1\right)\dfrac{dT}{T} - \dfrac{dC}{ART} = -m\,\dfrac{dV}{V}.$

We do not know the law of relation between C and T_1, that is, how to communicate to the air the heat released from the water and ice formed. We are forced to make a hypothesis which is not rigorously exact, but which is sufficiently approximate.

We will suppose that the transmission is proportioned to the fall of temperature, and therefore that

$$dC = -\mu \gamma\, dT$$

in which

$$\gamma = \frac{r_1' + 79}{T_1' - T_2}$$

whence we have;

$$\frac{1}{AR}(mc+\mu_1 c_1+\mu_1\gamma)\frac{dT}{T}=-m\frac{dV}{V}$$

integrating we get

$$\frac{c}{AR}\left(1+\frac{\mu_1 c_1+\mu_1\gamma}{mc}\right)l\frac{T'_1}{T_2}=l\frac{V_2}{V''_1},$$

or

$$\frac{1}{k-1}\left(1+\frac{\mu_1 c_1+\mu_1\gamma}{mc}\right)l\frac{T'_1}{T_2}=l\frac{V_2}{V''_1} \quad (29)$$

we have furthermore

$$P_0 V_2 = RmT_2,$$
$$P_1 V''_1 = RmT'_1,$$

whence

$$\frac{T'_1}{T_2}=\frac{P_1 V''_1}{P_0 V_2}.$$

Equation 29 can then be written;

$$\frac{1}{k-1}\left(k+\frac{\mu_1 c_1+\mu_1\gamma}{mc}\right)l\frac{T'_1}{T_2}=l\frac{P_1}{P_0}. \quad (30)$$

We can obtain the value of T_2 by successive approximations.

An approximate value for T_2 is found to be

$$T_2=\frac{\left\{mkc+\mu_1 c_1-\frac{(k-1)mc}{2}\frac{P_1}{lP_0}\right\}T'_1+\mu_1(r'_1+79)}{mkc+\mu_1 c_1+\frac{(k-1)mc}{2}l\frac{P_1}{P_0}}$$

Suppose now that condensation occurs in the cooler, we find by the tables the pressure of p_1' of saturated vapor of temperature T_1' and we can deduce the weight of the vapor condensed in the cooler.

We shall have then;
$$C = \mu_1 x_1'(r' + 79)$$
and
$$\gamma = x_1' \frac{r_1' + 79}{T_1' + T_2}$$

The equations 29 and 30 apply in this case as in the preceding.

The quantity of disposable negative heat is;
$$Q = mkc(T_0 - T_2) \qquad (31)$$
since we suppose the negative heat of the snow formed to be lost.

Finally the work produced by the expansion is;
$$W_m = P_1 V_1'' + \frac{mc}{A}\left(1 + \frac{\mu_1 c_1 + \mu_1 \gamma}{mc}\right)$$
$$(T_1' - T_2) - P_0 V_2 \quad (32)$$
or
$$W_m = \frac{mkc + \mu_1(c_1 + \gamma)}{A}(T_1' - T_2). \quad (33)$$

If there is a condensation in the cooler,

we should replace μ_1 in equations 32 and 33 by $\mu_1 x_1'$.

§ 14. The following table gives the cooling and general effect obtained from a cubic meter of air supposing a hygrometric state of $\frac{1}{2}$ and a temperature of 15°. The weight of the air is then $1.^k 2157$ instead of $1.^k 226$ which is the weight of dry air at this temperature.

We have also $p_\bullet = 85.^k 8$ and $\mu_1 = 0.^k 00626$.

Pressure atmospheres.	Temperature in compressor.	Temperature of expelled air.	Cooling obtained. $t_0 - t_s$.	Negative calories.	Theoretic work, kilogrammeters.	Effective work expended.	No. of negative calories obtained.				Weight of vapor carried into expansion cylinder.
							Per kilogrammeter, theoretic.	Per theoretic horse power per hour.	Per effective kilogrammeter.	Per effective horse power.	
	degrees	degrees	degrees	cal.			cal.	cal.	cal.	cal.	
1½	51,04	— 0,43	15,43	4.455	201	981	0,01531	4.134	0,00454	1.226	0,00626
2	79,31	—21,70	36,70	10.600	1.257	2.429	0,00843	2.276	0,00436	1.177	0,00626
2½	102,93	—37,25	52,25	15.090	2.286	3.843	0,00660	1.782	0,00393	1.061	0,00615
3	123,39	—50,80	65,80	18.997	3.486	5.354	0,00545	1.472	0,00355	959	0,00504
3½	141,57	—61,53	76,53	22.095	4.592	6.730	0,00481	1.299	0,00328	886	0,00439
4	157,98	—70,58	85,58	24.710	5.674	8.050	0,00435	1.176	0,00307	829	0,00384
4½	173,00	—78,26	93,26	26.928	6.714	9.304	0,00409	1.104	0,00290	783	0,00343

In comparing this table with the table of § 11 we see that the influence of the humidity of the air upon the results obtained is the greater when the pressure is low. We have made a similar remark in reference to the passive resistances. The theoretical advantage therefore of low pressures is practically much diminished by these causes of loss.

It is possible to neutralize almost completely the influence of moisture in the air. To accomplish this it would suffice to employ the air after it had produced its cooling effect and had parted with its moisture. It would be necessary to make the refrigerating machine a closed machine, making the same quantity of air serve indefinitely. The cooling would be produced by causing the cooled air to pass through an apparatus surrounded by some liquid not easily frozen, such as a solution of calcium or magnesium chloride. A part of the negative calories would thus be used, as well as by direct contact, and so many as are not used would not be lost, as the air passes directly to the compressor A, not at 15° as before, but $a-8°$ or $-10°$ of temperature. We think that it is only in this way that we can improve the air machine so that it can compare favorably with the machines using a liquefrable gas.

INFLUENCE OF WASTE SPACES.

§ 15. We will suppose the air to be dry in order to avoid complexity in our calculations.

Preserving our previous notation and calling v the amount of useless space in the compression cylinder, and v' that of the expansion cylinder; μ the weight of air enclosed in the space v at the end of the compression, we have;

$$P_0(V_0+v)^k = P_1(V_1+v)^k \qquad (34)$$
$$P_0(V_0+v) = R(m+\mu)T_0 \qquad (35)$$
$$P_1 V_1 = RmT_1 \qquad (36)$$
$$Pv_1 = R\mu T_1 \qquad (37)$$

m being the weight of dry air driven out of the compressor.

Equations (34), (35), (36) and (37) give by elimination of μ

$$T_1 = T_0 \left(\frac{V_0 + v}{V_1 + v}\right)^{k-1} \qquad (38)$$

and
$$T_1 = T_0 \left(\frac{P_1}{P_0}\right)^{\frac{k-1}{k}} \qquad (39)$$

The work of resistance to compression, taking account of the work restored to the piston as it begins to ascend, by the air in the waste space, expanding from P_1 to P_0, is;

$$W_r = \frac{k}{k-1}(P_1V_1 - P_0V_0)$$
$$+ \frac{1}{k-1} P_0 v \frac{V_0 - V_1}{V_1 + v}; \qquad (40)$$

For the cooling period;

$$V_1' = V_1 \frac{T_1'}{T_1} \qquad (41)$$

and
$$P_1 V_1' = RmT_1' \qquad (42)$$

The heat Q_1 absorbed by the water of the condenser is;

$$Q_1 = mkc(T_1 - T_1') \qquad (43)$$

PERIOD OF EXPANSION.—The air coming from the reservoir R' under pressure P_1 and the temperature P_1', should at the moment of opening of the inlet valve cause the air in the waste space and whose volume is v', to change its pressure from P_0 to P_1. This influences the temperature T_1'' of the mixture, also the weight m' of the air which passes from the reservoir into the waste space.

The dimensions of the reservoir being very large in comparison to the waste spaces, we may assume that no change occurs either in temperature or pressure of the reservoir, while the waste spaces are filled with air at the pressure P_1.

Calling μ' the weight of the air enclosed in the waste space at the moment that the inlet valve opens. We have;

$$P_0 v' = R\mu' T_2; \qquad (44)$$

T_2 being the final temperature of the expanded air.

The stored up work of this air is;

$$\frac{c}{A}\mu' T_2.$$

The weight m' of air filling the waste space, and having a temperature T_1' and a pressure P_1 has a stored energy of

$$\frac{c}{A} m' T_1'.$$

After the waste space is filled, the stored up energy of the total quantity of air $m' + \mu'$ contained there is

$$\frac{c}{A}(m' + \mu')T_1''$$

and we have furthermore;

$$P_1 v' = R(m' + \mu')T_1''. \qquad (45)$$

As we suppose there is neither loss nor gain of heat from the exterior, the difference between the stored energy of the mixture after the mass m' is introduced, and the sum of the stored energies of the masses m' and μ' before mixing is equal to the external work performed.

This exterior work is evidently $P_1 v_1'$, and calling the volume of m' before its introduction into the cylinder under pressure P_1 and temperature T_1' equal to v_1', then;

$$P_1 v_1' = RmT_1'.$$

We have also

$$-\frac{c}{A}\mu T_2 - \frac{c}{A}m' T_1' + \frac{c}{A}(m' + \mu')T_1'' = RmT_1',$$

Replacing $\dfrac{c}{A}$ by $\dfrac{R}{k-1}$ and combining with equations 44 and 45

$$m' = \frac{(P_1 - P_0)v'}{kRT'} \qquad (46)$$

and $\quad T_1'' = \dfrac{km'T_1' + \mu'T_2}{m' + \mu'}$

or $\quad T_1'' = \dfrac{kP_1 T_1' T_2}{P_1 T_2 + P_0(kT_1' - T_2)}$ $\qquad (47)$

When the inlet valve closes, the piston has described a volume V_1'', which has been filled by the weight m'' of air at pressure P_1 and temperature T_1'. We have then;

$$m' + m'' = m$$

There is no external work performed upon the total mass of air, since the negative work of the piston $P_1 V_1''$ is exactly equal to the positive work exerted by the air of the reservoir. The weights and temperatures of the air at the beginning and the end of the introduction possess the following relations:

$$\frac{c}{A}(m'+\mu)T_1'' + \frac{c}{A}m''T_1'$$
$$= \frac{c}{A}(m'+m''+\mu')T_1'''.$$

T_1''' being the temperature at the end of the introduction.

This equation gives;

$$\left. \begin{array}{l} T_1''' = \dfrac{(m'+\mu')T_1'' + m''T_1'}{m+\mu'} \\[2mm] \text{or} \\ T_1''' = \dfrac{P_1(V_1'+v') - \dfrac{1}{k}(P_1-P_0)v'}{P_1V_1'T_2 + P_0v'T_1'} T_1'T_2 \end{array} \right\} \quad (48)$$

we also have

$$P_1(V_1''+v') = R(m+\mu')T_1'''$$

or

$$P_1(V_1''+v') = P_1(V_1'+v')$$
$$- \frac{1}{k}(P_1-P_0)v'; \quad (49)$$

V_1' is given equations 38 and 41. Equation 49 gives the value of V_1''.

The inlet valve being closed, the mass of air $m+\mu$ which is at pressure P_1 and temperature T_1''' expands without gain or loss of heat since we neglect the influence of the sides of the cylinder. At the end of the stroke, this volume becomes V_2+v', its temperature T_2 and its pressure P_2. We have then

$$\left.\begin{array}{l} P_2(V_2+v')^k = P_1(V_1''+v')^k \\ \text{or} \\ P_2(V_2+v')^k = P_1 \left\{ (V_1'+v') - \dfrac{1}{k}\left(1-\dfrac{P_0}{P_1}\right)v' \right\}^k \end{array}\right\} \quad (50)$$

and
$$P_2(V_2+v') = R(m+\mu')T_2 \qquad (51)$$

Equations 50 and 51 give V_2 and T_2 if P_2 be known, or P_2 and T_2 if V_2 is known; this latter being the volume described by the piston of cylinder B.

We have
$$T_2 = T_1''' \left(\dfrac{P_2}{P_1}\right)^{\frac{k-1}{k}}$$

When there is no waste space we have
$$T_2 = T'_1 \left(\dfrac{P_2}{P_1}\right)^{\frac{k-1}{k}}$$

As T_1''' is greater than T_1', it results that for a given weight of air passed through the machine, at a given working pressure, that the final temperature of the expanded air would be higher, and consequently the number of negative calories produced would be less than if there had been no waste spaces.

The work is equal to:

$$W_m = \frac{k}{k-1}(P_1V_1' - P_2V_2) + (P_2 - P_0)V_2$$
$$+ \frac{1}{k-1}(P_0 - P_2)v' \quad (52)$$

§ 16. In order that the machine should work to the best advantage it is evidently necessary that the air should leave the cylinder at atmospheric pressure, that is, that P_2 should equal to P_0. There ought then to exist a certain relation between the volume of the compression cylinder $V_0 + v$, the pressure in the reservoir P_1 and the volume of the expansion cylinder $V_2 + v'$ which may be determined by the above equations. To fix the dimensions of a machine we may assume $V_0 + v$ and P_1 as given, and then deduce the value of $V_2 + v'$.

If we make $P_2 = P_1$ equations 50 and 51 will become

$$P_0(V_2+v')^k = P_1\left\{(V_1+v') - \frac{1}{k}\frac{P_1-P_0}{P_1}v'\right\}$$

and $\quad P_0 V_2 = RmT_2,$

whence
$$V_2 + v' = (V_0 + v)\left(\frac{P_0}{P_1}\right)^{\frac{k-1}{k}}$$
$$\left(\frac{T_1'}{T^0} = \frac{1}{k}\frac{P_1-P_0}{P_0} \cdot \frac{v'}{V_0+v}\right) \cdot (53)$$

The work is

$$\left.\begin{aligned} W_m &= \frac{k}{k-1}(P_1 V_1' - P_0 V_2) \\ \text{or} \quad W_m &= \frac{mkc}{A}(T_1' - T_2) \end{aligned}\right\} \quad (54)$$

This value for the work is the same as found in § 7, where no waste space was allowed for; only the final temperature T_2 being greater for the same weight and pressure, the work of the air is less.

The work of the resistance of the machine is then:

$$W_r - W_m = \frac{k}{k-1}[P_1(v_1-v_1')$$
$$-P_0(v_0-v_2)] + \frac{k}{k-1}P_0 v \frac{V_0-V_1}{V_1+v}$$

or $$W_r - W_m = \frac{mkc}{A}(T_1-T_1'-T_0+T_2)$$ (55)

The negative heat produced is

$$Q = mkc(T_0-T_2) \quad (56)$$
$$Q_1 - Q = W_r - W_m \quad (57)$$

The performance of the machine is

$$u = \frac{T_0-T_2}{T_1-T_1'-T_0+T_2}$$

or $$\frac{T_1-T_1'}{T_1-T_1'''} \cdot \frac{T_0}{T_1-T_0} \quad (58)$$

As T_1''' is greater than T_1, the useful effect is less than if there had been no waste space.

§ 17. The following table exhibits the results of a machine having waste space of 4 per cent. of the volume described by the pistons. The amount of air used being a cubic meter at 15°, and weighing

P_1 (Atmospheres.)	Temperature of outflow from compressor.	Temperature of final outflow.	Cooling of air.	Negative calories.	Theoretical work.	Effective work.	Negative calories developed.			
							Per theoretic kilogrammeter.	Per theoretic horse power per hour.	Per effective kilogrammeter.	Per effective horse power per hour.
	degrees	degrees	degrees	cal.			cal.	cal.	cal.	cal.
1½	51,04	—14,17	29,17	8,492	478	1.150	0,017777	4.800	0,007385	1.984
2	79,31	—34,76	49,76	14,486	1.426	2.582	0,010160	2.743	0,005611	1.515
2½	102,93	—49,17	64,17	18,682	2.563	4.094	0,007290	1.968	0,004563	1.231
3	123,39	—60,33	75,33	21,931	2.711	5.555	0,005910	1.596	0,003948	1.066
3½	141,57	—69,21	84,21	24,516	4.859	6.970	0,005045	1.362	0,003517	950
4	157,98	—76,57	91,57	26,658	5.976	8.321	0,004470	1.207	0,003204	864
4½	173,00	—82,78	97,78	28,468	7.063	9.617	0,004031	1.088	0,002960	799

1^k 226. In the cooler the air is brought to 18°.

By comparing these results with those of § 11, we see that the effect of waste spaces is by no means to be neglected since it results in a loss of about 100 calories for each theoretic horse power per hour.

§ 18. We can neutralize the influence of waste space by closing the outlet valve of cylinder B before the end of the stroke, so as to compress the air in this space; the stroke of the piston being exactly determined, the air in the waste space may be brought at the opening of the inlet valve to the temperature T_1' and the pressure P_1'.

In this case the equations 34 and 43 apply without change.

During the period of expansion we have:

$$P_2(V_2+v')^k = P_1(V_1'+v')^k \qquad (59)$$

$$P_2(V_2+v') = R(m+\mu')T_2 \qquad (60)$$

$$P_1 V_1 = R\mu' T_1' \qquad (61)$$

whence

$$\frac{T_2}{T_1'} = \left(\frac{V_2+v'}{V_1'+v'}\right)^{k-1} = \left(\frac{P_2}{P_1}\right)^{\frac{k-1}{k}} \quad (62)$$

The work restored by piston B is to make allowance for the compression of air in the waste space from the pressure P_0 to P_1:

$$\left.\begin{aligned} W_m &= \frac{k}{k-1}(P_1V_1' - P_2V_2) - \\ &\quad (P_0-P_2)\frac{k}{k-1}P_0v'\frac{V_0-V_1}{V_1+v} \\ &\quad + \frac{1}{k-1}(P_0-H_2)v' \end{aligned}\right\} \quad (63)$$

or

$$W_m = \frac{mkc}{A}(T_1' - T_2) + (k-1)\frac{mc}{A}T_2$$

$$\left(1 - \frac{P_0}{P_2}\right) - (k-1)\frac{\mu'c}{A}\left(\frac{P_0}{P_2}T_2 - T_2'\right);$$

T_2' being the temperature of the air in the cylinder at the moment compression commences before the end of the stroke.

We have then:

$$\left.\begin{aligned}W_m &= \frac{k}{k-1} \\ &[P_1(V_1-V_1')-P_0V_0+P_2V_2]+ \\ +(P_0&-P_2)V_2+\frac{k}{k-1}P_0(v-v') \\ &\frac{V_0-V_1}{V_1+v}-\frac{1}{k-1}(P_0-P_2)v'.\end{aligned}\right\} \quad (64)$$

When the machine is well regulated, the final pressure $P_2 = P_0$ and the equations 63 and 64 become

$$W_m = \frac{k}{k-1}(P_1V_1'-P_0V_2)$$

$$+\frac{k}{k-1}P_0v'\frac{V_0-V_1}{V_1-v} \quad (65)$$

or $\qquad W_m = \frac{mkc}{A}(T_1'-T_2)$

and

$$\left.\begin{aligned}W_r-W_m &= \frac{k}{k-1} \\ &[P_1(V_1-V_1')-P_0(V_0-V_2)]+ \\ +\frac{k}{k-1}\cdot P_0(v-v')&\frac{V_0-V_1}{V_1+v}.\end{aligned}\right\} \quad (66)$$

We have also:

$$\frac{V_0+v}{V+v} = \frac{V_2+v'}{V_1'+v'}. \quad (67)$$

We see that in equation 66 the term relating to waste spaces disappears if we make $v = v'$. The equation then becomes

$$W_r - W_m = \frac{k}{k-1}[P_1(V_1 - V_1') - P_0(V_0 - V_2)]$$

The volume $V_2 + v'$ is determined by means of equations 39, 41 and 67 when the pressure P_1 is known.

Reciprocally when V_0, v, V_2 and v' are known (the dimensions of the machine) then V_1' is readily found, and consequently P_1 and T_1, the pressure and temperature at the end of the stroke in cylinder B to insure the escape of the air at the atmospheric pressure.

§ 19. It was remarked in § 5 that the efficiency of the machine could be notably improved by cooling the air in the interior of the compressor cylinder.

This result can be accomplished, in part at least, if not completely, by means of a ject of water, such as is employed in compressed air engines.

We will proceed to calculate the work necessary for the compression in this particular case, neglecting the effect of waste spaces.

Let m be the weight of dry air occupying the volume V_0. Let M represent the weight of water injected together with the amount of moisture in the air, and Mx the weight of the vapor at any instant.

The dilatation or compression of the mixture of the vapor and air is effected in such manner as to satisfy the differential equation:

$$mcdt + \mathrm{M}(dq + dx\rho) = -\mathrm{A}Pd\mathrm{V}. \quad (69)$$

which expresses the fact that variations in the internal heat of the mixture equal the variations of work accomplished.

We have also

$$dq = c_1 dt,$$

c being the specific heat of water.

The differential equation can then be written

$$(mc + \mathrm{M}c_1)dt = -\mathrm{M}dx\rho$$
$$-\mathrm{A}(\mathrm{P}-p)d\mathrm{V} - \mathrm{A}pd\mathrm{V}$$

p being the tension of the vapor, and P that of the mixture.

But
$$x\rho = xr - Apxu$$
$$dx\rho = dxr - Apdxu - Axudp.$$
and
$$dV = Mdxu$$

ρ is the latent heat of the vapor,

r is the heat of vaporization,

u is the increase of volume of a kilogram of water vaporized.

We know furthermore that
$$Axu\frac{dp}{dt} = \frac{xr}{T}.$$

We have then
$$Mdx\rho + ApdV = Mdxr - \frac{xrdt}{T}$$
or
$$\frac{Mdx\rho + ApdV}{T} = Md\frac{xr}{T},$$

from which we deduce
$$(mc + Mc_1)\frac{dt}{T} + Arm\frac{dV}{V} = -Md\frac{xr}{T}.$$

Integrating between the limits T_1 and T_0,
$$(mc + Mc_1)l\frac{T_0}{T_1} + M\frac{x_0 r_0}{T_0} - M\frac{x_1 r_1}{T_1}$$
$$+ ARml\frac{V_0}{V_1} = 0, \qquad (70)$$

$$Mx_0 = \frac{V_0}{u_0} \text{ and } Mx_1 = \frac{V_1}{u_1};$$

$\frac{1}{u_0}$ and $\frac{1}{u}$ are very nearly the reciprocals of the vapor densities under the pressures p_0 and p_1.

We have furthermore

$$V_1 = Rm \frac{T_1}{P_1 - p_1}.$$

Equation 70 will give M when T_1 and T_0 are known.

$$AW_r = mc(T_1 - T_0) + M(q_1 - q_0) + x_1 \rho_1 - x_0 \rho_0) + A(P_1 V_1 - P_0 V_0)$$

or

$$AW_r = mkc(T_1 - T_0) + M(q_1 - q_0 + x_1 r_1 - x_0 r_0) \qquad (71)$$

This equation gives the work of resistance when M has become known.*

---

* The two equations 70 and 71, which express the relations between the volumes and the temperatures of a mixture of air and vapor, which is compressed or dilated, and which determine also the value of the work, are applicable to the Mekarski motor.

In this machine, which is designed to employ compressed air, the air is reheated just before it is introduced into the cylinder by being forced through water, having a temperature of 100° to 150°. The cylinder then contains air and saturated vapor, heated to a mean temperature of 100°.

In M. Colladon's compressors, into which a spray of water is injected, the air being compressed to four atmospheres, the temperature T_1 does not rise above 50° centigrade, the external air being about 50°.

We deduce then

$V_1 = 0.28429$ cu. metres

$M = 0.57212$

$W_r = 15.291$ kilogrammeters.

When the compression is effected without external cooling, we found in § 11 that the work of compression $= 17.649$ kilogrammeters, which shows a gain in the above process of about 13 per cent.

It remains to determine W_r for any pressure without any known value of T_1.

When a certain volume of air is dilated or compressed, with or without the addition of heat, the relation of pressure to volume is expressed by the equation

$$PV^a = \text{a constant.}$$

The weight M of equations 70 and 71 is then the weight of the vapor contained in air, saturated at the temperature at which it leaves the hot water.

$$\frac{V_1}{V_0} = \left(\frac{P_0 - p_0}{P_1 - p_1}\right)^{\frac{1}{a}} \qquad (72)$$

and
$$\frac{T_1}{T_0} = \left(\frac{P_1 - p_1}{P_0 - p_0}\right)^{\frac{a-1}{a}} \qquad (73)$$

which gives

$$\frac{a-1}{a} = \frac{\log T_1 - \log T_0}{\log(P_1 - p_1) - \log(P_0 - p_0)}; \qquad (74)$$

T_1 having been found by experiment, equation gives a.

Making 74 $P_1 = 4$ atmospheres, $T_1 = 323°$ and $T_0 = 288°$ we find $a = 1.0912$. a being thus determined equation 73 will give M_1. Only p_1 being a function of T_1, the latter must be found by successive approximations.

Equation 70 gives

$$Mc_1 = 0.4343 \frac{\frac{V_0 r_0}{u_0 T_0} - \frac{V_1 r_1}{u_1 T_1}}{\log \frac{T_1}{T_0}} + 0.5888\, m.$$

r_0, u_0, r_1 and u_2 are furnished by the tables.

Finally we obtain W_r by equation 71.

The saturated air in passing into the cooler is reduced in temperature from T_1 to T_1', and a portion of the vapor is condensed. The weight of vapor remaining and introduced into the expansion cylinder is:

$$\mu_1 = \frac{V_1'}{u_1'}$$

$\frac{1}{u_1'}$, being the density of the vapor corresponding to the temperature T_1'.

We will calculate again the cooling produce by the expansion and the work as explained in § 13.

§ 20. The following table exhibits the results obtained from a cubic meter of air saturated at 15°, since the sides of the compressor cylinder are covered with water. The weight of the air is $1^k\ 021$.

64

Pressures.	Temperature in compressor.	Temperature of air going out.	Reduction of Temperature.	Negative calories obtained.	Theoretic work developed.	Work of resistance.	Effective work developed.	Negative calories obtained.				Weight of vapor introduced into cylinder B.
								Per theoretic kilogrammeter.	Per theoretic H. P. per hour.	Per effective kilogrammeter.	Per effective H. P. per hour.	
	degrees	degrees	degrees	cal.				cal.	cal.	cal.	cal.	
1½	24,79	+ 5,79	9,21	2.636	189	4.266	856	0,0139	3.763	0,00308	832	0,01010
2	32,00	−18,62	33,62	9.623	709	6.687	1.833	0,01357	3.663	0,00525	1.417	0,00759
2½	37,68	−37,27	52,27	14.962	1.368	9.885	2.953	0,01094	2.953	0,00527	1.423	0,00609
3	42,38	−51,30	66,30	18.979	2.000	11.940	3.750	0,00949	2.562	0,00506	1.366	0,00509
3½	46,26	−62,48	77,48	22.179	2.636	13.723	4.621	0,00841	2.270	0,00480	1.296	0,00436
4	50,00	−71,60	86,60	24.790	3.244	15.291	5.431	0,00764	2.063	0,00456	1.231	0,00382
4½	53,13	−79,40	94,40	27.022	3.809	16.672	6.172	0,00709	1.914	0,00438	1.182	0,00340

An examination of this table and a comparison with the table of § 14 shows:

1st. That the injection of water into the interior of the compressor cylinder increases the efficiency 40 to 50 per cent.

2d. That the efficiency is at a maximum at a pressure of $2\frac{1}{2}$ atmospheres.

3d. That it diminishes, though slowly, as we vary from this pressure.

4th. That the quantity of snow or ice produced is not greater than that which comes from the moisture of the atmosphere.

The most favorable working pressure apppears to be in this case nearly 4 atmospheres, since we obtain then a sufficiently good result (24 to 25 negative calories for a cubic meter of air), with a relatively good performance of 1,200 negative calories per horse-power per hour.

Theoretically the injection of water into the compressor affords a great advantage. But it is possible that the water resulting from the condensation of vapor in the cooler does not all remain in the reservoir, but that a portion is carried mechanically into cylinder B.

The results indicated above for the efficiency would in such a case be considerably modified, and the increase in the quantity of frozen vapor would constitute in practice a grave inconvenience.

Experiment can alone decide this question.

We have examined in the preceding pages nearly all the problems belonging to the air machine. We will pass now to the study of the second class of machines, or those which transform motive force into negative heat by the employment of a liquefiable gas.

§ 21. THE principle of these machines is the same as that of the kind described in the last chapter. The gas is compressed, then deprived of its heat, and finally caused to expand in such a manner as to lower its temperature. Only in this instance the abstraction of the heat which follows the compression, has the effect to liquefy the gas, and it is the vaporization of the resulting liquid which produces the lowering of the temperature.

When a change of volume of a saturated vapor is made under constant pressure, the temperature remains constant. The addition or subtraction of heat, which produces the change of volume, is represented by an increase or a diminution of the quantity of liquid mixed with the vapor.

On the other hand when vapors, even if saturated, are no longer in contact with their liquids, and receive an addition of heat, either through compression by a mechanical force, or from some external source of heat, they comport themselves nearly in the same way as permanent gases, and become superheated.

It results from this property, that refrigerating machines, using a liquefiable gas will afford results differing according to the method of working, and depending upon the state of the gas, whether it remains constantly saturated, or is superheated during a part of the cycle of working.

§ 22. We will suppose first that the

gas is constantly saturated and will examine the conditions to be fulfilled under this hypothesis, and the results that may be obtained.

Employing the notation of the preceding chapter we will designate by m the weight of the gas employed, P_2 and T_2, the pressure and the absolute temperature of the cooled gas, P_1 and T_1', the pressure and the absolute temperature in the condenser.

The pressures P_2 and P_1 are determined by the temperatures T_2 and T_1'; These are the pressures of a saturated vapor at these temperatures, and are given in Regnault's tables.

The temperature of the condenser is determined beforehand by local conditions. Depending on the surface, the interior of the condenser will exceed by 5° or 10° the temperature of the water furnished to the exterior. This latter will vary from 11° or 12° C the temperature of water from considerable depth below the surface, to 30° or 35°, the temperature of surface water in hot climates. The volatile liquid employed in the machine ought not at this temperature to have a tension above that which can be readily managed by the apparatus.

On the other hand if the tension of the gas at the minimum temperature is too low, it becomes necessary to give to the compression cylinder large dimensions, in order that the weight of vapor afforded by a single stroke of the piston shall be sufficient to produce a notably useful effect.

These two conditions, to which may be added others; such as those depending on the greater or less facility of obtaining the liquid, upon the dangers incurred in its use either from its inflammability or unhealthfulness, and finally upon its action upon the metals, limit the choice to a small number of substances.

The gases or vapors in use, are; Sulphuric Ether, Sulphurous Oxide, Ammonia and Methylic Ether.

The following table, derived from

Regnault exhibits the tensions of the vapors of these four substances at different temperatures between $-30°$ and $+40°$. The original tables expressed the tensions in millimeters of mercury. To facilitate computation, the tensions are here given in *kilograms per square meter*.

Temperature.	Sulpuric Ether.	Sulphur Dioxide.	Ammonia.	Methylic Ether.
$-40°$	—	—	7.187	—
-35	—	—	9.302	—
-30	—	3.908	11.918	7.837
-25	—	5.082	15.120	9.736
-20	917	6.519	19.003	11.992
-15	1.194	8.265	23.669	14.652
-10	1.541	10.366	29.225	17.765
-5	1.968	12.874	35.797	21.380
0	2.493	15.840	43.475	25.547
$+5$	3.129	19.322	52.405	30.318
$+10$	3.894	23.378	62.707	35.743
$+15$	4.808	28.074	74.504	41.873
$+20$	5.891	33.474	87.925	48.755
$+25$	7.164	39.645	103.073	56.437
$+30$	8.651	46.659	120.083	64.961
$+35$	10.377	54.585	129.054	—
$+40$	12.367	63.496	160.112	—

An inspection of the table shows at once that the use of ether does not readily lead to the production of low temperatures because its pressure becomes then very feeble.

The ether machine is, however, abandoned. Ammonia on the contrary is well adapted to the production of low temperatures; but its elastic force is very great at temperatures from 15° to 30° which are readily produced in the condenser. It is not a good aid to the transformation of mechanical force into heat, on account of the difficulty of maintaining tight joints in the apparatus, and of the influence of waste spaces at the high pressures. Methylic ether yields low temperatures without attaining too great pressures at the temperature of the condenser. Finally, sulphur dioxide readily affords temperatures of $-10°$ to $-15°$ while its pressure is only 3 to 4 atmospheres at the ordinary temperature of the condenser. These two latter substances then lend themselves conveniently for the production of cold by means of mechanical force.

§ 23. Let c be the specific heat of the liquid employed.

q the quantity of heat necessary to raise 1 kilogram of the liquid from 0° to T°−273°.

$$q = c(T-273)$$

λ, r, ρ, the total heat, the heat of vaporization, and the latent heat of the vapor considered at the temperature T°−273.

u, the increase of volume of one kilogram of liquid vaporizing at T°−273°.

We have by definition

$$\lambda = r + q$$
$$\rho = r = AP u.$$

We will apply indices to these quantities similar to those which affect the letter T in designating the different absolute temperatures.

In order that the vapor be constantly saturated, it is necessary that the quantities of liquid and of vapor taken into the compressor at once be such that at the end of the compression all the liquid shall be vaporized and the vapor shall not be superheated.

If we let x'_2, represent the proportion of vapor contained in the mixture at the commencement of the inflow, the work of compression will be equal to the difference in the amount of internal heat of the mixture at the beginning and end of the compression, that is to say to $m(q_1' - q_2 + \rho_1' - x_2' \rho_2)$.

The work of the inflow into the condenser will be $P_1 V_1$, calling V_1' the volume occupied by a weight m of the vapor at the end of the compression, and the work of the back pressure will be $P_2 V_2$, V_2 being the volume occupied by the weight mx_2' of vapor.

We have also

$$V_1' = m\left(u'_1 + \frac{0{,}001}{\delta}\right)$$

and $$V_2 = mx'_2\left(u_2 + \frac{0{,}001}{\delta}\right), \quad (75)$$

δ being the density of the liquid supposed constant.

We may neglect the fraction which is very small, and write

$$V'_1 = mu'_1,$$
$$V_2 = mx'_2 u_2,$$

from which we may get

$$mr'_1 = m\rho'_1 + AP_1 V'_1$$

and
$$mr_2 = m\rho_2 + AP_2 V_2.$$

The total work of the compression including the outflow is

$$AW_r = m(q'_1 - q_2 + r'_1 - x'_2 r_2). \quad (76)$$

As the compression follows an adiabatic curve, the quantities q'_1, q_2, r'_1, r_2, T'_1 and T_2 bear the following relation:

$$\int_{T_2}^{T'_1} \frac{c\,dt}{T} = \frac{x'_2 r_2}{T_2} - \frac{r'_1}{T'_1}$$

or more simply,

$$\frac{r'_1}{T'_1} + cl\frac{T'_1}{T_2} = \frac{x'_2 r_2}{T_2}. \quad (77)$$

Equation (77) will give the quantity x'_2. Consequently equation (75) furnishes, when we know m, the volume V_2 that the piston should describe during the aspiration in order that all the liquid should be vaporized at the end of the compression; or, inversely, the weight m may be found if V_2 be given.

The vapor flows into the condenser where it is liquefied.

The heat absorbed by the water of the condenser is

$$Q_1 = mr'_1 \qquad (78)$$

The liquid, then passes into the expansion cylinder where it is vaporized, producing work till it attains the pressure P_2 and the temperature T_2 of the refrigerant. At the end of the expansion, the weight of vapor in the mixture is mx_2.

The work, including the counter-pressure, and neglecting the work of introducing the liquid, $p_1 \dfrac{0.00.1.m}{\delta}$, which is very small, is;

$$AW_m = m(q'_1 - q_2 - x_2 r_2). \qquad (79)$$

and the equation of the adiabatic curve is

$$\frac{x_2 r_2}{T_2} = cl \frac{T'_1}{T_2} \qquad (80)$$

which determines x'_2.

The quantity of heat Q necessary to bring the mixture whose weight is

$m(1-x_2)$ of liquid and mx_2 of vapor to its primitive condition, in which $m(1-x'_2)$ is the weight of the liquid and mx'_2 is the weight of the vapor, is,

$$Q = m(x'_2 - x_2)r_2$$

or by reason of equations (76) and (79)

$$Q = \frac{T_2}{T'_1} mr'_1 \qquad (81)$$

The work expended is $W_r - W_m$ and we have

$$A(W_r - W_m) = m[r'_1 - (x'_2 - x_2)r_2] = Q_1 - Q. \qquad (82)$$

The theoretic performance of the machine is

$$\frac{Q}{W_r - W_m} = \frac{AQ}{Q_1 - Q} = A \cdot \frac{T_2}{T'_1 - T_2} \qquad (83)$$

a result already found in section 3, and which is identical with that at which we arrived in the case of permanent or non-liquefiable gases.

§ 24. We will now take a numerical example, and consider the dimensions of the cylinders to be so regulated that a final temperature of $-15°$ is obtained, the temperature of the condenser being

+18°, and the volume of gas taken into the compressor at each stroke, $V_2 =$ one cubic meter.

The resolution of the above equations supposes a knowledge of the values of r, q, c and u, or APu. They have been determined directly by Regnault for sulphuric ether, but not for sulphur dioxide, ammonia and methylic ether. Availing ourselves of the experiments of Regnault upon the compressibility of gases, we have been able to determine these quantities for sulphur dioxide and ammonia and prepare tables giving results for every five degrees from $-30°$ to $+40°$.

The method of calculation of these tables will be found in a note at the end of this essay.

For sulphur dioxide we find,

$t_2 = -15$ or $T_2 = 258$ $t'_1 = +18$ or $T_2 = 291$
$r_2 = \ \ 95.015$ $r'_1 = \ \ 87.23$
$AP_2 u_2 = 7.932$ $AP_1 u'_1 = 8.568$
$q_2 = -5.4615$ $q'_1 = 6.554$
$u_2 = \ \ 0.419$ $u'_1 = 0.1165$

The table of § 22 gives $P_2=8265$ and $P_1=31170$.

Making the calculations indicated by the equations (77) and (80) we find

$x'_2 = 93.29$ per cent.
$x_2 = 11.90$ per cent.

Equation (75) gives

$m = 2.554$ kilograms.

Equations (76) and (79) give

$AW_r = 27.08$ whence $W_r = 11.482$ k'g'm.
$AW_m = 1.82$ whence $W_m = 772$ k'g'm.

Finally equations (78) and (81) give

$Q_1 = 222.77$
$Q = 197.56$

Thus the volume described by the piston of the compression cylinder being one cubic meter, $2^k,554$ of sulphur dioxide working between $-15°$ and $+18°$ produce 197.50 negative calories. To effect this it is necessary to introduce into the compressor cylinder at each stroke a mixture of liquid and gas of which the proportion should be 93.29 per cent. of gas and 6.71 per cent. of liquid.

. We have for ammonia

$$t_2 = -15° \qquad t'_1 = +18°$$
$$P_2 = 23669 \qquad P'_1 = 82183$$
$$r_2 = 322.53 \qquad r'_1 = 301.70$$
$$AP_2 u_2 = 28.604 \qquad AP_1 u'_1 = 31.431$$
$$u_2 = 0.512 \qquad u'_1 = 0.1621$$
$$q_2 = -14.68 \qquad q'_1 = 18.696$$

The mean specific heat of the liquid at $0°$, $c = 1.0058$.

By means of these given values we find

$x'_2 = 92.62$ per cent.
$x'_2 = 9.68$ per cent.
$m = 2^k.1034$
$AW_r = 76.55 \qquad W_r = 32,457\,\text{k'g'm}$
$AW_m = 4.52 \qquad W_m = 1,917\,\text{k'g'm}$
$Q_1 = 634.59$
$Q = 562.56$

$2^k.1034$ of ammonia working between the same limits of $+18°$ and $-1°$ and with the same dimensions of compressor cylinder as before furnish 562.56 negative calories per hour.

We will now consider ether. The vapor of ether, unlike steam, superheats during expansion and condenses during compression. An ether machine ought, therefore, to work so that only vapor is introduced into the compressor cylinder, and not a mixture of liquid and vapor. At the end of the compression a part of the vapor becomes condensed.

We shall then have $x'_2 = 1$ and the equations above found become:

$$V'_1 = m x'_1 \left(u'_1 + \frac{0{,}001}{\delta} \right),$$

$$V'_2 = m \left(u_2 + \frac{0{,}001}{\delta} \right),$$

$$\frac{x'_1 r'_2}{T_2} = \frac{r_2}{T_2} - c \cdot l \frac{T'_1}{T_2},$$

$$\frac{x_2 r_2}{T_2} = cl \frac{T'_1}{T_2},$$

$$Q_1 = m x'_1 r'_1,$$

$$Q = m(1 - x_2) r_2,$$

$$AW_r = m(q'_1 - q_2 + x'_1 r'_1 - r_2),$$

$$AW_m = m(q'_1 - q_2 - x_2 r_2).$$

The empirical formulas established by Regnault for the vapor of ether are:

$$r = 94{,}00 - 0{,}0790 t - 0{,}0008514 t^2,$$
$$APu = 7{,}46 + 0{,}02747 t - 0{,}001354 t^2,$$
$$q = 0{,}52901 t + 0{,}0002959 t^2.$$

and we deduce:

for $t = -15$ and $t = +18$,
$P_2 = 1194$ kilog. $P_1 = 5456$,
$r_2 = 94963$, $r_1 = 92{,}302$,
$AP_2 u_2 = 7{,}014$, $AP_1 u_1 = 7{,}516$,
$u_2 = 2{,}491$,
$q_2 = -7.868$, $q_1 = 9{,}618$,
$c = 0{,}5299$,

and we have $\delta = 0{,}736$.

Performing the calculations indicated, we find,

$x_2 = 17.35$,
$x'_2 = 100$,
$x'_1 = 95.64$,
$m = 0.401$,
$Q_1 = 35°.40$,
$Q = 31°.38$,
$AW_r = 4.44$, $W_r = 1882^{kgm}$.
$AW_m = 0.42$, $W_m = 178^{kgm}$.

The same machine working between $-15°$ and $+18°$, will give per cubic meter of—

Ammonia...... 562.56 negative calories.
Sulphur dioxide. 197.56   "
Sulphuric ether. 31.38   "

The efficiency would be 0°,0184 per kilogrammeter.

§ 25. We remark here that the positive work W_m is always small compared with the negative work W_r.

We can then without great loss of power simplify the machine by suppressing the expansion cylinder and replacing it by a simple cock so regulated as to deliver into the cooler a quantity of liquid precisely equal to the amount admitted to the compressor to obtain the determined cooling effect.

The cycle of operations is not reversible. We shall have $\dfrac{Q}{AW_r} = \dfrac{Q}{Q_1 - Q}$, but the proportion $\dfrac{Q}{Q_1 - Q}$ will be less than $\dfrac{T_2}{T'_1 - T_2}$, and the efficiency would be less.

This manner of working is represented in the diagram, Fig. 1, by replacing the adiabatic line $V'_1 V_2$ by the two right lines $V'_1 V'''_2$ and $V'''_2 V'_2$, situated to the right of the point V_2. The quantity Q proportioned to $V''_2 V_0$ is less than the

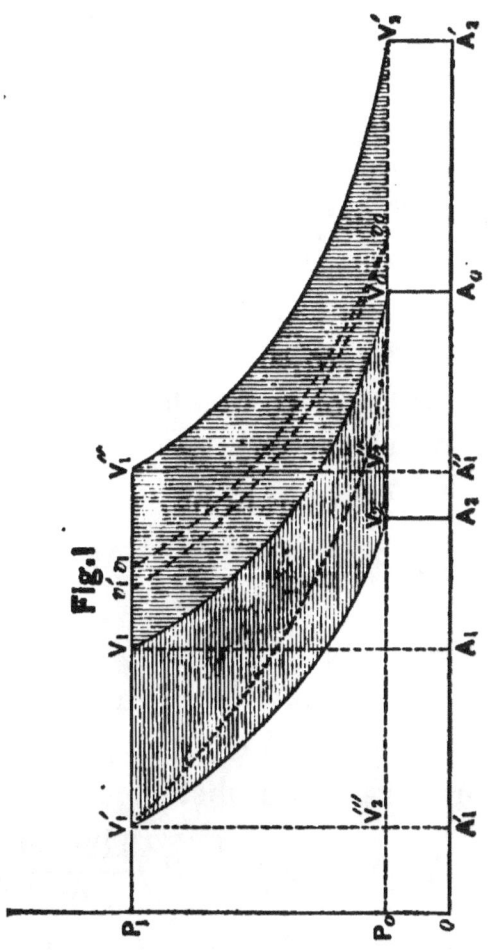

quantity Q of the preceding case which was proportioned to $V_2'V_0$, and the quantity Q_1-Q will be augmented by a quantity proportional to the area $V'_1V'''_2V_2$.

The equations (76), (77) and (78) remain unchanged.

The weight m of the liquid under the pressure P_1 and the temperature T'_1, passing suddenly into the refrigerator, a part of the liquid is vaporized; the temperature of the mixture becomes T_2 and the pressure P_2. The quantity x_2 of liquid, which is vaporized, is given by the equation

$$m(q_2-q'_1+x_2\rho_2)+AP_2V'_2-A(P_1-P_2)\frac{0.001.m}{\delta}=0,$$

which shows that the variation of internal heat $m(q_2-q'_1+x_2\rho_2)$ is equal to the exterior work accomplished;

$$-AP_2V'_2+A(P_1-P_2)\frac{0.001.m}{\delta};$$

V'_2 being the volume occupied by the weight mx_2 of vapor after the passage of the mixture into the refrigerant.

We have $V'_2 = mx_2\left(u_2 + \dfrac{0.001}{\delta}\right)$.

If we neglect the very small quantity
$$AP_1 \dfrac{0.001\, m}{\delta}$$
the preceding equation becomes:
$$x_2 r = q'_1 - q_2 \qquad (84)$$

The quantity Q is again given by the equation
$$Q = m(x'_2 - x_2)r_2$$
or by reason of eq. (76)
$$Q = mr'_1 - AW_r = Q_1 - AW_r$$
from whence the performance
$$\dfrac{Q}{W_r} = A\dfrac{Q}{Q_1 - Q} \qquad (85)$$

The efficiency will be less. It is easy to show that the value of x_2 given by eq. (84) is always greater than that given by eq. (80). Consequently the value of Q will be less in the second case than in the first, and the ratio $\dfrac{Q}{Q_1 - Q}$ will also be less.

In applying equations (84) and (85) to the same cases as those of § 24, we find for sulphur dioxide

$$x_2 = 12.64 \text{ per cent.}$$
$$Q = 195.71$$

and the performance $= 0.^c 0170$ per kilogrammeter. For ammonia:

$$x_2 = 10.35 \text{ per cent.}$$
$$Q = 558.11$$

and the efficiency $0.^c 0172$.

Finally for sulphuric ether

$$x_2 = 18.46 \text{ per cent.}$$
$$Q = 30.96$$

efficiency $= 0.^c 0164$

§ 26. In order to realize, either the cycle of Carnot or the non-reversible cycle indicated above, it is necessary, when we employ a liquefiable gas which superheats under compression, to introduce into the compressor cylinder at each aspiration, a mixture of liquid and vapor in such proportions that it shall all be in the state of gas at the end of the compression.

We can devise no practical means of realizing this condition. So we content ourselves when employing freezing machines that use a liquefiable gas, with introducing into the compressor the gas without any mixture of liquid. It happens then with sulphur dioxide and ammonia that the gas superheats during compression, and therefore that during a part of the operation the machine acts like the air machine.

It is clear that under these conditions we augment the range of temperature between T_1 of the gas arriving in the condenser, and T_2 of the refrigerant, and consequently of the useful effect of the apparatus.

Referring again to Fig. 1 we see that we start with a volume v_0 greater than V_0 of the preceding case, compress the vapor to the volume v_1 following the adiabatic curve $v_0 v_1$ of the superheated gas; cool it from the temperature T_1 to the temperature T_1' corresponding to its liquefaction under the pressure P_1. It is then passed into the refrigerant either producing work and describing the adiabatic curve $V_1 V_2$ or by means of a cock by which means it describes the lines $V_1' V_2'''$ and $V_2''' V_2''$.

The quantity of negative heat gained

by superheating is represented by the length $V_0 v_0$ and the increase of resistant work by the area $V_0 V_1 v_1 v_0$.

Tracing from the point v_0 the adiabatic curve of the saturated vapor, the point v_1' will be to the left of v_1.

If the compressed vapor follows the adiabatic $v_0 v_1'$, the performance $\dfrac{Q'}{Q_1'-Q}$ will be equal to the performance $\dfrac{Q}{Q_1-Q}$ of the cycle $V_0 V_1 V_1' V_2$.

But as the compression follows the line $v_0 v_1$ we see that for the same quantity Q' of obtainable negative heat, the quantity $Q_1 - Q$ would be greater than a quantity proportional to the area $v_0 v v_1'$.

We can say, then, that *a priori*, the theoretic efficiency of freezing machines working so as to superheat the gas is less than that of machines that work without superheating.

The difference is small as we shall see later.

§ 27. We will now examine the conditions of working of a machine, under the supposition that we introduce into the cylinder during aspiration only gas, and in such condition as to superheat during compression.

A certain volume V_2 of gas under pressure P_2 and temperature T_2, it is required to find its volume V_1 and its temperature T_1 when it shall have attained the pressure P_1 of the condenser.

If liquefiable gases behaved as do permanent gases, it would suffice to use the equations (1) to (6), which were established in § 10 for the compression of air.

But the researches of Regnault on the compressibility of gases, have established the fact that when near the liquefying point these bodies are far from following the laws of Mariotte and Gay Lussac upon which the formulas which we have used were founded.

Zeuner has given (Théorie Mécanique de la Chaleur) the result of his researches upon superheated steam.

He found the following relation to exist between the pressure P, the volume of the unit of weight (specific volume) v, and the absolute temperature T,

$$P_v = BT - CP^n \qquad (86)$$

in which C and n are constants to be determined by experiment

$$B = \frac{C_p n}{A} \qquad (87)$$

C_p being the specific heat of the vapor under constant pressure, which is constant according to Regnault.

If we make $k=\frac{1}{4}$, $B=50.933$ and $C=192.50$, we find that this formula furnishes for the specific volume of steam, numbers which agree remarkably well with the results of experiment.

Zeuner does not offer this relation as rigorously exact, but as giving much better results than the formula,

$Pv = RT$ which applies to permanent gases.

Liquefiable gases being nothing but superheated vapors, we will employ equation (84) established for superheated steam, but will determine the constants in each case employing the results of Regnault's experiments upon the dilatation and compression of gases.

If we call a the coefficient of dilatation of the gas under atmospheric pressure, it is easy to see that eq. (86) gives:

$$a = \frac{1}{273 - \frac{C}{B} \cdot 10.334^n}$$

and $\quad 10.334 v_0 = 273 B - C \cdot 10334^n, \quad$ (88)

whence $\quad 10.334 v_0 a = B. \quad$ (89)

an equation which gives B when we know the coefficient of dilatation and specific volume v_0 at $0°$ and atmospheric pressure.

If the relation (87) were exact, it would suffice with equations (88) and (89) for determining B, C and n. But the numbers thus obtained do not coincide, at least in the case of sulphur dioxide and ammonia with the results obtained by Regnault. Instead therefore of using equation (87) we will determine n by one of the results found by Regnault for the product PV.

Regnault gives values of PV for temperatures of $1.7°$ for sulphur dioxide, for $8.1°$ for ammonia and for pressures varying from 600 to 1200 and 1400 millimeters of mercury. We can deduce from these tables the volume V_0 at 0° and under pressure of 760 millimeters, and then calculate the weight m of the gas required in our examples. We then have

$$P\cdot V = mBT - mCP^n \qquad (90)$$

which combined with equation (89) will furnish C and n.

For sulphur dioxide

$$a = 0.0039028; \quad v_0 = 0.3442$$

For $P = 16.345^{kgm.}$ and $T = 274.7$.

Regnault found,

$$\frac{PV}{m} = 3526.16$$

We deduce $\quad B = 13.882$

$$C = 3.8455$$

$$n = 0.44487$$

Introducing these constants into equation (86) we can obtain for Pv values which coincide in a satisfactory manner with Regnault's results.

These values are slightly less than Regnault's for pressures between 10.334 kg. and 16.345 kg., and a little larger for pressures lying beyond these limits on either side.

For ammonia we unfortunately do not know the coefficient of dilatation; it was not determined by Regnault. As this gas is near its liquefying point at 0° we will assume its coefficient to be about the same as that of sulphur dioxide and cyanogen, which is 0.0039. In the absence of exact values determined by experiment it is clear that results obtained under the above assumption can be regarded as approximative only.

We have $v_0 = 1.2977$ and Regnault's tables give:

$$\frac{PV}{m} = 13596 \text{ for } T = 281.1° \text{ and } P = 19515 \text{ kgm.}$$

We then deduce

$B = 52.4943$, $C = 43.7144$, $n = 0.32685$,

§ 28. It remains now to find the equation of the adiabatic curve of a superheated vapor, of which the pressure, the specific volume and the temperature are related as follows:

$$pv = BT - Cp^n.$$

The fundamental equation of the mechanical theory of heat is, calling Q the quantity of heat furnished to a body, U its internal work, and supposing the external pressure is always equal to the expansive force:

$$dQ = A(dU + pdv),$$

and as U is a function of p and v, we have:

$$dQ = A \left\{ \frac{dU}{dP} dp + \left(\frac{dU}{dv} + p \right) dv \right\}.$$

Assuming $\dfrac{dU}{dp} = X \quad \dfrac{dU}{dv} + p = Y,$

we have $\quad dQ = A(Xdp + Ydv) \quad (91)$

or $\quad dQ = AT \left(\dfrac{x}{T} dp + \dfrac{Y}{T} dv \right)$

and since dU is an exact differential,

$$\frac{dX}{dv} = \frac{dY}{dp} - 1.$$

We know that the factor $\dfrac{1}{T}$ is the factor of integrability of the function $Xdp + Ydv$; and we deduce

$$T = Y\frac{dt}{dp} \times \frac{dt}{dv} \qquad (92)$$

We also have in virtue of equation (86),

$$\frac{dt}{dp} = \frac{v}{B} + \frac{nCp^{n-1}}{B}$$

and
$$\frac{dt}{dv} = \frac{p}{B}.$$

If we suppose that the pressure remain constant, $dp=0$, and eq. (91) gives

$$dQ_p = AYdv.$$

But, $dQ_p = c_p dt$, calling c_p the specific heat at constant pressure, which we suppose constant and which is known. We have then:

$$Y = \frac{c_p}{A}\frac{dt}{dv} = \frac{c_p}{A}$$

and from eq. (92),

$$X = -\frac{BT}{p} + \frac{c_p}{AB}(v + nCp^{n-1})$$

and finally,

$$dQ = A$$
$$\left\{ \frac{c_p}{AB}(pdv + vdp) - vdp + \left(\frac{nc_p}{AB} - 1\right)Cp^{n-1}dp \right\}$$
$$(93)$$

For the equation of an adiabatic curve, it is necessary to make $dQ=0$. We have then:

$$\left(\frac{c_p}{AB}-1\right)dpv + pdv + \left(\frac{nc_p}{AB}-1\right)Cp^{n-1}dp = 0. \tag{94}$$

Introducing the value of T from equation (86), it becomes.

$\dfrac{c_p}{AB}\dfrac{dt}{T} = \dfrac{dp}{p}$ and integrating $\dfrac{c_p}{AB}lT = lp + \text{const.}$

or finally

$$\left(\frac{T}{T_0}\right)^{\frac{c_p}{AB}} = \frac{p}{p_0}, \tag{95}$$

an equation analogous to equation (4), which we found for air.

Replacing T by this value in equation (86) we get finally for the equation of the adiabatic curve

$$pv = BT_0\left(\frac{p}{p_0}\right)^{\frac{AB}{c_p}} - CP^n. \tag{96}$$

If $\dfrac{AB}{c_p}$ be equal to n, as Zeuner admits, for superheated steam, this equation becomes $pv_k=$ a constant, and it is similar to that which represents the adiabatic curve of the permanent gases.

Eq. (94) gives the work of compression

$$pdv = -d\mathrm{W}\left(1-\frac{c_p}{\mathrm{AB}}\right)dpv + \left(1-\frac{nc_p}{\mathrm{AB}}\right)\mathrm{C}p^{n-1}dp$$

whence

$$\mathrm{W} = \left(\frac{c_p}{\mathrm{AB}}-1\right)(pv-p_0v_0) + \left(\frac{c_p}{\mathrm{AB}} - \frac{n}{1}\right)\mathrm{C}(p^n-p_0^n) \quad (97)$$

or again

$$\mathrm{W} = \left(\frac{c_p}{\mathrm{AB}}-1\right)\mathrm{B}(\mathrm{T}-\mathrm{T}_0) + \mathrm{C}\left(1-\frac{1}{n}\right)(p^n-p_0^n) \quad (98)$$

and

$$\mathrm{W} = \left(\frac{c_p}{\mathrm{AB}}-1\right)\mathrm{B}(\mathrm{T}-\mathrm{T}_0) + \mathrm{C}\frac{1-\frac{1}{n}}{p_0^n}\left\{\left(\frac{\mathrm{T}}{\mathrm{T}_0}\right)^{\frac{nc_p}{\mathrm{AB}}}-1\right\} \quad (99)$$

§ 29. We can now establish the equations relating to the compression of a liquefiable gas in a cylinder. A weight m of gas occupying the volume V_2 at the temperature T_2, and under the pressure P_2 is compressed until the pressure is P_1 of the condenser. The temperature T_1

at the end of the compression will be given by the equation (95).

$$T_1 = T_2 \left(\frac{P_1}{P_2}\right)^{\frac{AB}{c_p}} \quad (100)$$

and the work of compression including the flowing of the the gas is

$$W_r = \frac{mc_p}{AB}(P_1V_1 - P_2V_2) + m\left(\frac{AB}{c_p} - \frac{1}{n}\right) \cdot \frac{C(P_1^n - P_2^n)}{n}$$

or

$$W_r = \frac{mc_p}{A}(T_1 - T_2) - \frac{mcP_2^n}{n}\left\{\left(\frac{T_1}{T_2}\right)^{\frac{nc_p}{AB}} - 1\right\} \quad (101)$$

m is given by the equation

$$= \frac{P_2 V_2}{BT_2 - CP_2^n} = \frac{V_2}{u_2 + \frac{0.001}{\delta}} \quad (102)$$

the final volume

$$V_1 = V_2 \frac{P_2}{P_1} \cdot \frac{BT_1 - CP_1^n}{BT_2 - CP_2^n}$$

We cool the gas in the condenser under constant pressure. The volume V_1 becomes V_1' at the moment the temperature becomes T_1'; since the gas is liquefied we have;

$$V_1' = V_2 \frac{P_2}{P_1} \cdot \frac{BT_1' - CP_1^n}{BT_2 - CP_2^n}$$

and the quantity of heat removed from the condenser is:

$$Q_1 = mc_p(T_1 - T_1') + mr_1' \qquad (103)$$

The volume occupied by the liquid is

$$v_1 = \frac{0.001.m}{\delta}$$

δ being the density of the liquid supposed constant.

The liquid is then passed into the refrigerant without producing work.

The quantity mx_2 of gas which vaporizes while the pressure passes from P_1 to P_2 and the temperature from T_1' to T_2 is by equation (84);

$$mx_2 r_2 = m(q_1' - q_2).$$

The quantity of negative heat obtained is:

$$Q = m(1 - x_2)r_2$$
or $$Q = m(\lambda_2 - q_1') \qquad (104)$$

and we have

$$Q_1 - Q = mc_p(T_1 - T_1') + m(r_1' + q_1' - r_2 - q_2)$$
or $$Q_1 - Q = mc_p(T_1 - T_1') + m(\lambda_1' - \lambda_2)$$

We can verify the equality $Q_1 - Q = AW_r$ or

$$\lambda_1' - \lambda_2 = c_p(T_1' - T_2) - \frac{AC}{n}(P_1{}^n - P_2{}^n)$$

Referring to the fundamental equation

$$dQ = AdU + APdv$$

and making $dQ = 0$ it becomes

$$mdU = -mPdv = -dW.$$

and consequently

$$U_1 - U_2 = \left(\frac{C_p}{AB} - 1\right)B(T_1 - T_2) + C\left(1 - \frac{1}{n}\right)(P_1^n - P_2^n) \quad (105)$$

We have furthermore by definition,

$$\lambda = AU + APv,$$

an equation which signifies that the total heat of the vapor at $t°$ is equal to the internal heat AU augmented by the thermal equivalent of the work of vaporization and dilatation.

We have then

$$\lambda_1 - \lambda_2 = C_p(T_1 - T_2) - \frac{AC}{n}(P_1^n - P_2^n)$$

This equation is applicable to a superheated vapor above its point of saturation.

It applies also at the point of saturation; we have then

$$\lambda_1' - \lambda_2 = C_p(T_1' - T_2) - \frac{AC}{n}(P_1^n - P_2^n) \quad (106)$$

which verifies the equation

$$Q_1 - Q = AW_r.$$

Equation (105) can be written:

$$U - U_0 = \left(\frac{C_p}{AB} - 1\right)(Pv - P_0v_0) + \left(\frac{C_p}{AB} - \frac{1}{n}\right)C(P^n - P_0^n)$$

If we make $\frac{C_p}{AB} - \frac{1}{n} = 0$.

the equation becomes

$$U = U_0 + \frac{1-n}{n}(Pv - P_0v_0)$$

Under this form it expresses Hirn's law of superheated vapors, and may be thus expressed:—from the point of condensation, to the point at which the superheated vapor possesses the same properties as the permanent gases, the product pv remains constant while the internal work remains the same.

But the equation

$$\frac{c_p}{AB} = \frac{1}{n}$$

is not verified for the cases of the two liquefiable gases which we have studied, and consequently we cannot apply to them the law of Hirn.

§ 30. We will now take a numerical example and suppose as in the preceding case, that a cubic meter of gas is admitted at the temperature of $-15°$ under a pressure corresponding to this temperature, and that it is compressed until its tension is that of the condenser and that the temperature of this latter is, in the interior, $+18°$.

Sulphur dioxide. Equation (102) gives

$$m = \frac{1}{0.419 + \frac{0.001}{1.42}} = 2^k.382$$

Equation (100) gives; making

$$c_p = 0.15438$$

after Regnault, and

$$\frac{AB}{c_p} = 0.211882;$$

$$T_1 = T_2 \left(\frac{P_1}{P_2}\right)^{0.211882} = 334.31 \quad \text{or} \quad t_1 = 68°.80$$

Equations (103) and (104) give
$$Q_1 = 227.49$$
$$Q = 197.75$$
whence $AW_r = Q_1 - Q = 28.71$
and $W_r = 121.75$

and the theoretic performance $= 0.^c0162$ or 4.374 calories per horse power per hour.

In a double acting engine working at high velocity we estimate the resistances at about 15 per cent. of the power expended.

$1.15 W_r = 13.998$ and the performance becomes $0.^c0141$ or 3.807 calories per horse power, per hour.

This performance is double that of the machine working with dry air between the same limits of temperature. This difference shows not that the air is theoretically a less efficient agent in the production of cold, but that to produce the same useful effect, the air machine having much larger dimensions than the liquefiable gas machines will experience proportionally greater loss through resistances.

§ 31. Generally with sulphur dioxide

we do not get as low a temperature as
$-15°$.

The opening of the cock which leads from the condenser to the cooler is so regulated that the pressure in the latter is about $\frac{9}{10}$ of an atmosphere, which corresponds to a temperature of $-12°.41$.

$$P_2 = 9301 \text{ kg.} \quad t_2 = -12°.41.$$

With these values the tables, given at the end of this memoir, give

$$m = \frac{1}{v_2} = 2^k.784$$
$$r_2 = 94.377$$
$$q_2 = -4.517$$
$$u_2 = 0.3863$$

and by means of equations (100), (102), (103) and (104) of § 29 it is easy to calculate T_1, Q_1, Q and W_r.

The results of these calculations are recorded in the following table, which gives the negative heat obtained, the work absorbed and the performance per cubic meter of sulphur dioxide, supposing the apparatus regulated for a temperature of $-12°41$ in the refrigerant, and that the temperature of the interior of the condenser varies from $+15°$ to $+40°$:

Temperature of condenser, inside.	Pressure corresponding P_1	Temperature after compression, t_1.	Heat absorbed by water of the condenser.	Negative heat.	Theoretical work of compression.	Work including passive resistances.	Negative calories developed.			
							Per theoretic kilogrammeter.	Per theoretic horse power per hour.	Per effective kilogrammeter.	Per effective horse power per hour.
degrees	kilog.	degrees	cal.	cal.	kgm.	kgm.	cal.	cal.	cal.	cal.
15	28,074	56.20	243.35	218.09	10,735	12,335	0.02031	5.484	0.01768	4.774
20	33,474	68.69	243.19	213.38	12,444	14,310	0.01715	4.630	0.01491	4.026
25	39,645	81.15	243.04	208.66	14,609	16,890	0.01428	3.856	0.01242	3.353
30	46,659	93.56	242.80	203.94	16,515	19,022	0.01235	3.334	0.01072	2.894
35	54,585	105.93	242.56	199.21	18,425	21,189	0.01081	2.919	0.00940	2.538
40	63,496	118.26	242.27	194.48	20,312	23,359	0.00957	2.584	0.00832	2.246

We see that the performance diminishes more than one-half when the temperature of the interior of the condenser rises from 15° to 40°.

The figures of the last column do not nearly represent the number of calories really produced and utilized. It is necessary to take into account the loss occasioned by the pipes; the waste spaces in the cylinder; of loss of time in opening of the valves; of the leakage around the piston and valves; of the reheating by the external air; and finally, when ice is being made, of the quantity of the ice melted in removing the blocks from their molds.

It requires about 100 calories to congeal to $-7°$ a kilogram of water taken at 15° or 16°. Manufacturers estimate that practically the sulphur dioxide apparatus using water at 12° or 13°, produces 25 kilograms of ice, or 2,500 calories per horse power per hour, measured on the driving shaft, which is about 55 per cent. of the theoretic efficiency indicated above.

Fig. 3 represents the Pictet machine from a design furnished us by the inventor. It has a double-acting compression cylinder with four valves. The cylinder is furnished with a jacket, within which a current of cold water is made to circulate.

The gas is compressed to a tension corresponding to the temperature of the water employed for cooling, generally 1.8 to 2 kilograms effective pressure; then it is discharged by the pipe T into the condenser C where it is liquefied.

This condenser is like the surface condensers of marine engines. It has a surface of about 24 square meters for 100,000 *theoretic* calories per hour, or 48 square meters for 100,000 *effective* calories per hour measured by the ice produced.

The quantity of water employed depends upon the difference of temperature to be allowed between the inside and outside of the condenser.

If this difference is to be 5° each litre of water releases 5 calories and the

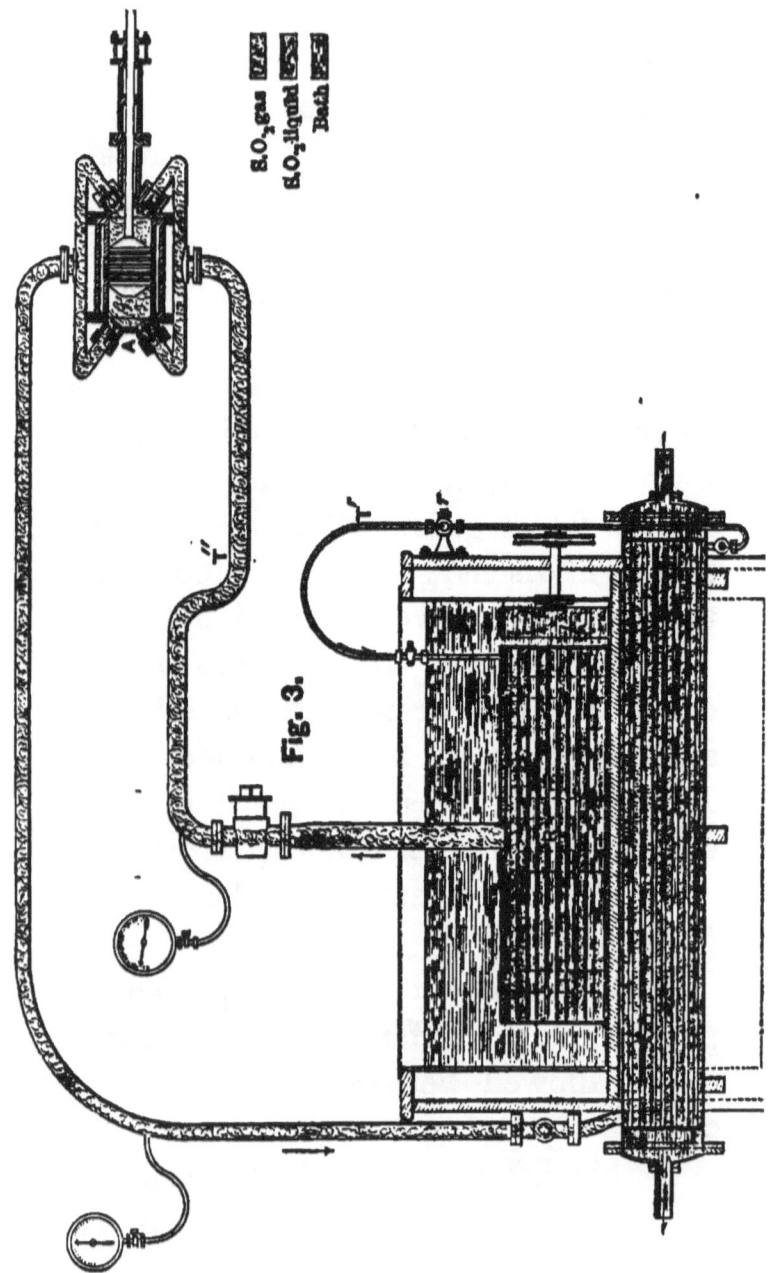

Fig. 3.

quantity of water to be employed will be for 100 theoretic calories produced

$$\frac{Q_1 100}{5Q} = 20\frac{Q_1}{Q}$$

which would require for the example of § 31 and for a temperature of 20° in the condenser, 22.8 litres.

The liquid dioxide passes into the refrigerant R by the pipe T', the supply being regulated by the cock r so that the pressure shall be $\frac{9}{10}$ of an atmosphere in the refrigerant and 3 atmospheres in the condenser. If the outlet by the cock be diminished the pressure is lowered in the cooler, and the temperature is also lowered, but the useful effect also diminishes, since for the same volume described by the compressor piston, less weight of gas is used. We have in this machine, therefore, the same facilities for varying the useful effect as in the air machines.

The refrigerant is constructed like the condenser. Its surface is 29 square meters for each 100,000 theoretic negative calories produced per hour. It is immersed in an incongealable bath formed of a solution of calcium chloride.

The temperature of the interior of the refrigerant being $-12°$, that of the bath being $-7°$. In this bath are immersed the tanks or moulds within which the water is frozen.

Finally the sulphur dioxide returns to the compressor cylinder by the pipe T''.

The dioxide may be employed continuously so long as no air is permitted to enter the joints. Any leakage might lead to the production of the trioxide and possibly sulphuric acid which would lead to injury to machine. Exceptional care is required in maintaining tight joints.

Some experiments with an ammonia machine have not yielded very good resuls; but the want of success seems to have resulted rather from an imperfect action of the surface of the refrigerant than from any inherent defect in the gas itself. Ammonia gas prevents the advantage of affording about three times the useful effect as sulphur dioxide for the same volume described by the piston.

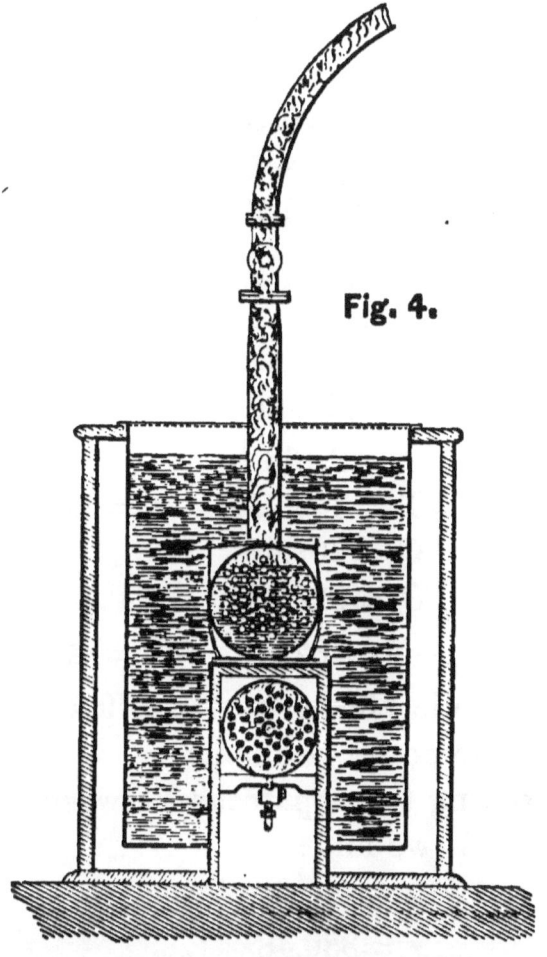

Fig. 4.

But this advantage is balanced by the inconvenience of higher pressures and consequently more leakage, &c.

Between the limits of temperature of 12°.41 in the refrigerant and + 18° in the condenser we find for ammonia:

$$P_2 = 26559 \text{ kilos.}$$
$$r_2 = 321.06$$
$$u_2 = 0.461$$
$$q_2 = -1219$$
and we have $C_p = 0.50836$
$$\frac{AB}{C_p} = 0.242615.$$

We deduce for each cubic meter described by the piston:

$$m = 2.163 \text{ k.}$$
$$T_1 = 342.75 \qquad t_1 = 69°.75$$
$$Q_1 = 709.48$$
$$Q = 627.03$$

$AW_r = Q_1 - Q = 82.45 \qquad W_r = 34.959$ kg.

Theoretic efficiency: 0.0179 or 4.833 per horse power per hour.

Working the apparatus between $-30°$ and $+18°$ we find.

$$P_2 = 11918k$$
$$r_2 = 330.48$$
$$u_2 = 0.9463$$
$$q_2 = -31.82$$
$$m = 1.0553k$$
$$T_1 = 388.20 \qquad t_1 = 115.20$$
$$Q_1 = 370.52$$
$$Q = 295.44$$

$AW_r = Q_1 - Q = 75.08 \qquad W_r = 31.834$

Theoretic result: 0.00928 or 2505 per horse power per hour.

Chapter IV.

MACHINES EMPLOYING CHEMICAL ACTION.

§ 34. It remains to discuss the ice making machines which employ chemical affinity in their mode of action, and of which the ammonia machine of M. Carré is the type.

Fig. 5 exhibits the disposition of the parts of this apparatus. It consists of a boiler A which contains a concentrated solution of ammonia in water; this boiler is heated either directly by a fire as shown in the figure, or indirectly by pipes leading from a steam boiler. The condenser B communicates with the upper part of the boiler by the tube *aa*; it is cooled externally by a current of cold water. The *réfrigérant* C is so constructed as to utilize the cold produced; the upper part of it is in communication with the lower part of the condenser by

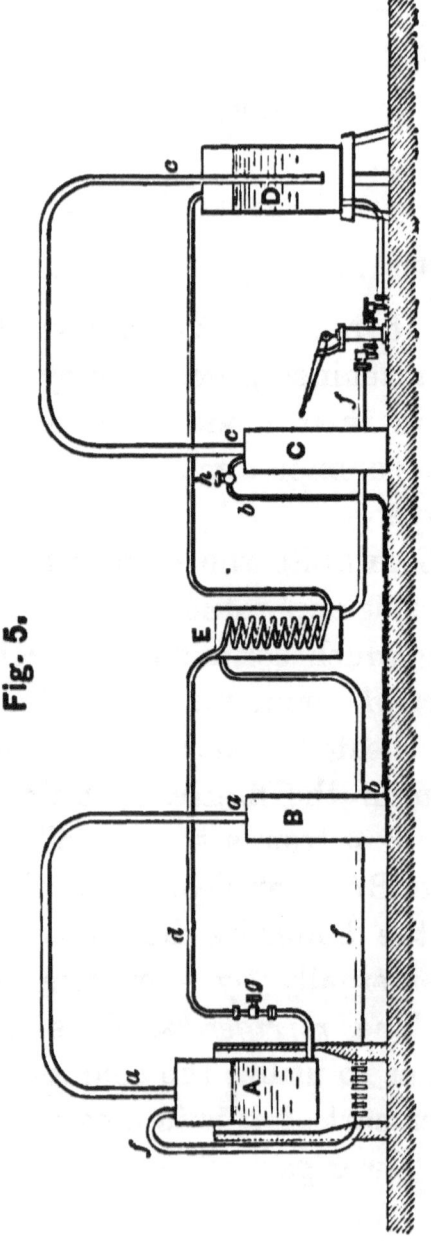

Fig. 5.

means of the tube bb. The details of the construction are not shown in the figure. An absorption chamber D is filled with a weak solution of ammonia; the tube cc puts this chamber in communication with the refrigerant C.

The absorption chamber communicates with the boiler by two tubes. One dd, leads from the bottom of the boiler to the top of the chamber D; the other, ff, leads from the bottom of D to the top of the boiler. Upon the pipes ff is mounted a little pump whose use is to force the liquid from the absorption chamber where the pressure is maintained at about one atmosphere, into the boiler, where the pressure is from 8 to 12 atmospheres.

The change of temperature is managed through the attachments to the pipes ff and dd in a manner that will be easily comprehended by an inspection of the figure.

To work the apparatus the ammonia solution in the boiler is first heated. This releases the gas from the solution and the pressure rises. When it reaches the tension of the saturated gas at the temperature of the condenser, there is a liquefaction of the gas, and also of a small amount of steam. By means of the cock h, the flow of the liquefied gas into the refrigerant C is regulated. It is here vaporized by absorbing the heat from the substance placed here to be cooled. As fast as it is vaporized it is absorbed by the weak solution in D. The small quantity of watery vapor is carried along mechanically.

Under the influence of the heat in the boiler A, the solution is unequally saturated, the stronger solution being uppermost.

The weaker portion is conveyed by the pipe dd into the chamber D, the flow being regulated by the cock g, while the pump sends an equal quantity of strong solution from D back to the boiler. While these exchanges are brought about in the solutions, there is also an exchange of temperatures whereby the weak liquid arrives cold in the absorption chamber, and the strong solution is delivered in the boiler hot.

The working of the apparatus depends upon the adjustment and regulation of the cocks g and h, and of the pump; by means of these, the pressure is varied, and consequently the temperature in the refrigerant C controlled.

It is seen that the working is similar to that of the machines described in the preceding chapters. The chamber D fills the office of aspirator, and the boiler A plays the part of compressor.

The mechanical force producing exhaustion, is here replaced by the affinity of water for ammonia gas; and the mechanical force required for compression is replaced by the heat which severs this affinity and sets the gas at liberty. We see then in advance that we shall again find a greater part of the equations already established in the discussion of the liquefiable gas machines.

§ 35. We will assume at first, that under the influence of the heat applied to the boiler, ammonia gas only is driven off, and no steam. We will assume a certain weight of the gas to enter the boiler in a state of solution; being heated, it will be separated from the water, requiring a certain quantity of heat which we will call Q'. Then, being conducted to the condenser, it will be cooled and then liquefied, and will impart to the water surrounding the coils a quantity of heat Q_1. In the refrigerant it is evaporated, borrowing from the exterior a quantity of heat Q; it is next absorbed by the liquid in the chamber D, disengaging a certain amount of heat to the liquid (which may be deducted from the total amount required in the boiler); and, finally, it is reconveyed to the boiler, where it arrives in its original condition. By reason of the exchange of temperature effected at E, all the heat of the weak solution going out of the boiler, is restored to the strong solution entering it, so that the changes of temperature in the water are effected without expenditure of heat.

In the complete cycle if we neglect the

small amount of work performed by the pump, and the heating and cooling due to contact with the air, it is clear that all the heat from external sources, being Q' from the boiler, and Q from the refrigerant, will be equal to the amount Q_1 carried away by the water of the condenser.

We have then

$$Q' = Q_1 - Q \qquad \text{and the}$$

efficiency will be expressed by

$$\frac{Q}{Q_1 - Q} \qquad \text{which is identical}$$

with that found for the machines depending on mechanical action.

Q' the quantity of heat which it is necessary to expend in order to produce the quantity Q of negative calories, being equal to $Q_1 - Q$, has the same value as the quantity AW_r, the calorific equivalent of the mechanical work expended in the machines previously discussed, to produce this same quantity Q of negative calories. We proceed to show that between the same limits of temperature in the condenser and refrigerant, and for the same value of Q, the quantity Q′ in this class of machines, is equal, very approximately at least, to the quantity AW_r.

We arrive then at this remarkable result; that in all the ice machines, when they work between the same limits of temperature, the theoretic quantity of negative heat produced is exactly the same for each calorie expended, whether it is directly produced by chemical action, or indirectly under the form of mechanical work.

But as a calorie represented by 424 kilogrammeters costs in the best heat motors an expenditure of at least 10 calories in the fire, it would seem that the chemical machines possess a considerable advantage over all the others, since in these latter the heat is employed directly, and not under the expensive form of mechanical work. Practically, however, this advantage is much less than that which seems to result from the above calculations; as we will proceed to show.

§ 36. We will assume the hypothesis mentioned in the beginning of the preceding section, and determine the quantities Q', Q_1 and Q in terms of the temperatures, the pressures and weights of the gas employed.

We will preserve the notations of the previous chapter. T_1 being the absolute temperature of the gas as it enters the condenser; T_1' its absolute temperature in the condenser, and T_2 the absolute temperature in the refrigerant.

Let m be the weight of the gas considered, occupying the volume V_2 at the temperature T_2, and under the pressure P_2 at its entrance into the absorption chamber.

Let AU be the internal heat at the temperature T; qe the heat necessary to raise a kilogram of water from $0°$ to $1°$.

After the gas has been absorbed by the water, the absolute temperature of the mixture will be T'_2.

During the process of absorption of the gas, there is an amount of external work accomplished equal to $P_2(V_2 - w)$, w of being the volume of water.

The difference in internal heat before and after this operation is equal to this external work. We have then

$$q_{e_2}' + mAU_2' - q_{e_2} - mAU_2 = AP_2(V_2 - w).$$

The solution is conveyed to the boiler, and there heated until all the gas is driven off. It then occupies the volume V_1 under the pressure P_1, and at the temperature T_1.

The necessary quantity of heat Q'' is equal to the difference in quantities of internal heat, augmented by the exterior work accomplished. This work is equal to $P_1(V_1-w)$ less the work of the pump, $(P_1-P_2)w$.

We have then

$$Q'' = q_{e_1} - q_{e_2}' + mAU_1 - mAU_2 + AP_1(V_1-w) - A(P_1-P_2)w.$$

Adding this equation to the preceding, member to member, we find

$$Q'' = q_{e_1} - q_{e_2} + mA(U_1-U_2) + AP_1V_1 - AP_2V_2$$

This equation is established without taking account of the effect of exchange of temperature. There is furnished to the solution which enters the boiler a quantity of heat precisely equal to $q_{e_1}-q_{e_2}$. The quantity of heat Q' to be supplied by the boiler, in order to bring the pressure of the gas from P_2 to P_1, and from the temperature T_2 to T_1 is then

$$Q'=mA(U_1-U_2)+AP_1V_1-AP_2V_2 \quad (107)$$

The equations 101 and 105 gave, in case of compression by a mechanical force,

$$AW_r=mA(U_1-U_2)+AP_1V_1-AP_2V_2$$

which is identical with the preceding.

We have then $Q'=AW_r$ provided that the temperature T_1 in the case where the change of pressure of the gas is obtained by the heat combined with the chemical action, is the same as in the case where the change is due to a mechanical force. Experiment proves that it is nearly so.

It appears that the temperature to which it is necessary to heat the ammonia solution to obtain a given pressure is higher as the solution becomes weak. Now in the ice machines the solution conveyed to the boiler contains rather less of the gas as the pressure in the refrigerant becomes more feeble. We understand therefore how the temperature T_1 ought to increase as the temperature T_2 of the refrigerant diminishes. Unfortunately, precise experiments upon this point are wanting.

A series of observations made by M. Rouart upon a Carré machine is herewith given.

The first column of each table gives the absolute pressures in atmospheres and kilograms; the second the temperatures observed in the boiler; the fourth, the temperatures of water in the condenser; the fifth column gives the temperatures of the liquefied gas corresponding to the pressures in the first column (see table in § 22); the temperatures are those of the interior of the condenser, and are naturally more elevated than the exterior.

In the case of mechanical compression the final temperature T_1 is related to the initial temperature and to the initial and final pressures as expressed by the equation (100)

$$T_1 = T_2 \left(\frac{P_1}{P_2}\right)^{\frac{AB}{Cp}}$$

The third column of the table gives the temperatures calculated by this formula, supposing $T_2 = 243$ and $P_2 = 11,918$.

For the mean pressures the calculated temperatures coincide nearly with the observations. For the higher pressures the calculated pressures are higher than the observed. But it is necessary to remark that in this case the watery vapor mixed with the gas exerts a greater influence, and that the true gas pressures ought to be sensibly less than the pressures which have served as a basis for calculation.

The condensation in the condenser and the evaporation in the refrigerant, are brought about exactly as in the case of the machines acting by mechanical force. We shall have then, as in § 27,

$$Q_1 = mc_p(T_1 - T_1') + mr_1',$$
$$x_2 r_2 = q_1' - q_2,$$

Pressure in Boiler.		Temperature of Boiler	
Atm.	Kilog.	Observed.	Calculated
		Degrees.	Degrees.
1½	15,501	48	—
2	20,668	58	—
2½	25,835	65	—
3	31,002	70	—
3½	36,169	75	—
4	41,336	80	—
4½	46,503	84	—
5	51,670	88	—
5½	56,837	92	—
6	62,004	94	—
6½	67,171	100	—
7¼	74,921	106	106
7½	77,505	108	109
8	82,672	112	116
8½	87,839	116	121
9	93,006	120	127
9½	98,173	124	132
10	103,340	128	137
10½	108,507	132	142
11	113,674	136	147
12	123,998	142	156
13	134,332	146	164
14	144,666	152	172
15	155,000	156	180
15	155,000	158	180

Temperature of water of condenser.	Temperature inside of condenser.	Difference.	Remarks.
Degrees.	Degrees.	Degrees.	
9	—	—	
9	—	—	
9	—	—	
9	—	—	
9	—	—	
9	—	—	
9	—	—	
9	—	—	
9	—	—	
9	—	—	
10	—	—	
10	15.0	5.0	The gas liquefies and the apparatus begins to work.
10	16.1	6.1	
12	18.0	6.0	
13	20.0	7.0	
14	21.7	7.7	
15	23.3	8.8	
17	25.1	8.1	
17	26.7	9.7	
19	28.1	9.1	
24	31.0	7.1	
30	32.8	2.8	
35	36.0	1.0	
37	38.0	1.0	
39	38.0	—	

Absolute pressures.		Temperature of Boiler	
Atm.	Kilog.	Observed.	Calculated
		Degrees.	Degrees.
3	31,002	73	—
4½	46,503	90	—
5	51,670	94	—
5½	56,837	100	—
6	62,004	103	—
6½	67,171	106	—
7	72,338	110	—
7½	77,505	118	109
8	82,672	124	116
8½	87,839	130	121
9	93,006	136	127
9½	98,173	140	132
10	103,340	146	137
10¾	111,089	147	145
11¾	121,423	148	153
13	134,332	148	164
14	144,666	154	172
15	155,000	160	180
15½	160,167	163	180

$$Q = m(1-x_2)r_2 = mr_2 - m(q_1' - q_2),$$

$$m = \frac{V_2}{u_2 + \frac{0.001}{\delta}} = \frac{V_2}{v_2}$$

$$Q' = Q_1 - Q.$$

Temperature of condenser water (observed).	Temperature of interior of condenser (calculat'd)	Difference of temperatures.	Observations.
Degrees.	Degrees.	Degrees.	
8	—	—	
8	—	—	
8	—	—	
8	—	—	
8	—	—	
8	—	—	
9	14.1	—	⎫ The liquefied
10	16.1	6.1	⎬ gas appears.
12	18.0	6.0	
14.5	20.0	5.5	
15	21.7	6.7	
16	23.3	7.3	
17	25.1	8.1	
19	27.4	8.4	
16 (?)	30.4	14.4 (?)	
16 (?)	32.8	16.8 (?)	
35	36.0	1	
38	38.0	0	
38	40.0	1	

The two following tables give the results of calculations for one cubic meter of ammonia gas, for temperatures in the condenser ranging from $+15°$ to $+40°$. In the first the temperature of the interior of the refrigerant is taken at −15°. In the second table it is −30°.

The numbers in the last column are calculated on the supposition that a kilogram of coal burned yields 4000 calories.

First case: $t_2 = -15°$, $m = 1^k.932$.

Temp. interior of condenser.	Temperature of boiler.	Calories removed by condenser.	Negative calories produced.	Calories expended.	Theoretic performance.	Theoretic result per kilogram of coal.
deg.	deg.	cal.	cal.	cal.	cal.	cal.
15	67.77	638.71	564.83	73.88	7,645	30,580
20	81.74	640.76	554.49	86.27	6,427	25,708
25	95.69	642.61	543.98	98.63	5,515	22,060
30	109.61	644.12	533.29	110.83	4,813	19,252
35	123.47	645.53	522.43	123.10	4,244	16,976
40	137.27	646.57	511.31	135.26	3,779	15,116

Second case: $t_2 = -30°$, $m = 1^k.023$.

Temp. interior of condenser.	Temperature of boiler.	Calories removed by condenser.	Negative calories produced.	Calories expended.	Theoretic performance.	Theoretic result per kilogram of coal.
deg.	deg.	cal.	cal.	cal.	cal.	cal.
15	106.07	361.77	293.04	68.73	4,263	17,052
20	121.61	364.09	287.57	76.52	3,771	15,084
25	137.13	366.31	282.01	84.30	3,345	13,380
30	152.61	368.38	276.35	92.03	3,003	12,012
35	168.03	370.31	270.60	99.71	2,714	10,856
40	183.38	372.09	264.71	107.38	2,465	9,860

The results indicated by the preceding tables are large; they vary from 9,860 to 30,580 negative calories for each kilogram of coal burned. We are far from attaining such results in practice.

We have omitted in our calculations to take into account two conditions which modify largely the theoretical results:

1st. The necessity of cooling the absorption chamber so that the solution of the gas may be readily accomplished.

2d. The influence of the water carried along with the gas.

We will now examine the influence of these two causes of loss.

§ 37. When ammonia gas dissolves in water, considerable heat is disengaged.

M. M. Fabre and Silbermann have measured this heat of solution, and found it equal to $514.^{cal}3$ for each kilogram of gas dissolved.

The liquid of the absorption chamber being employed continually in dissolving the gas from the refrigerant, rises rapidly in temperature, and as the solubility diminishes with the temperature, it soon reaches a condition at which it ceases to work. To insure successful working it is necessary, therefore, to treat the absorption chamber to a current of cold water in such a manner as to maintain a constant temperature. We will suppose this to be the same as that of the condenser t_1'.

If we denote by Q_1' the quantity of heat, of which the absorption chamber is relieved, we shall evidently have

$$Q_1 + Q_1' = Q + Q'$$
or
$$Q' = Q_1 + Q_1' - Q.$$

On the other hand, the gas arriving at the condenser, is always mixed with a certain quantity of steam, usually about 6 or 8 per cent. By employing a solution of calcium chloride instead of pure water for a solvent, the amount of watery vapor is reduced to about three per cent.

The presence of the steam reduces the efficiency to a notable extent. It carries off a portion of the heat of the boiler, and, having arrived in the refrigerant, it does not evaporate, but, by holding a portion of the ammonia, prevents it from volatilizing. It impedes the action then, nearly in the same way as the waste spaces in the mechanical action machines, but to a greater extent.

We will proceed to determine the influence of this introduction of water.

Let m, as before, be the weight of gas sent out from the boiler; μ the weight of water accompanying it, and the quantities r and q affected by the index e, shall relate to the water.

When the mixture passes into the condenser, the steam becomes liquid, and absorbs a certain weight m' of gas; and we have

$$m' = \frac{0.001 \beta_1' \mu}{v_1} \qquad (108)$$

β_1' being the coefficient of solubility by volume of the gas in water whose temperature is t_1'; $\frac{1}{v'}$ being the weight of a cubic meter of gas at this temperature.

According to Carius, the coefficient of solubility of ammonia, a gas in water, is represented by the empirical formula

$\beta = 1049.624 - 29.4963 t + 0.676873 t_2$
$\qquad - 0.0095621 t^3.$

The quantity of heat Q_1 which will be absorbed by the condenser, is equal to the quantity of heat necessary to lower the temperature of the weight m of gas from t_1 to t_1', plus the quantity of heat necessary to liquefy the weight $m - m'$ of gas, plus the quantity of heat necessary to liquefy, and raise to the temperature t_1' the weight μ of steam, plus the heat disengaged by the solution of the weight m' of gas.

We shall have then

$$Q_1 = mc_p(T_1 - T_1') + (m-m')r_1' + \mu (q_{e_1} - q_{e_1}' + r_{e_1}) + m's_1', \quad (109)$$

calling s_1' the heat disengaged by the solution of one kilogram of ammonia gas in water having the temperature t_1'.

The mixture passing into the refrigerant, a certain quantity of the liquefied gas is volatilized until the pressure and temperature become equal, respectively, to P_2 and T_2, the pressure and temperature of the refrigerant. The water will retain in solution a weight m'' of gas, given by the equation,

$$m'' = \frac{0.001 \beta_2 \mu}{v_2} \quad (110)$$

The quantity of gas volatilized $(m-m')x_2$ is found by the equation

$$\left. \begin{array}{l} (m-m')x_2 r_2 + (m'-m'') \\ \qquad s_2 = (m-m')(q_1 - q_2) + \\ + \mu(q_{e_1}' - q_{e_2}) + m'c_p(T_1' - T_2) \end{array} \right\} \quad (111)$$

The quantity of negative head realized is

$$Q = (m-m')(1-x_2)r_2 \qquad (112)$$

or

$$\left.\begin{array}{l}Q=(m-m')(r_2-q_1'+q_2)(m'-m'')-s_2\\ -\mu(q_{e_1}'-q_{e_2})-m'c_p(T_1'-T_2)\end{array}\right\} (113)$$

The quantity of heat Q_1' which it is necessary to supply to the absorption chamber in order to maintain a constant temperature, is equal to the heat arising from the solution of $m-m''$ weight of gas, minus the heat necessary to raise the weight m of gas and the weight μ of water, from T_2 to T_1'.

$$\begin{array}{r}Q_1'=(m-m'')s_1'-mc_p(T_1'-T_2)\\ -\mu(q_{e_1}'-q_{e_2}) \quad (114)\end{array}$$

The quantity of heat Q' which it is necessary to employ at the boiler, is equal to $Q_1+Q_1'-Q$. We have then, applying the above values,

$$\begin{array}{l}Q'=m[c_p(T_1-T_1')+s_1']+(m-m')\\ \qquad [\lambda_1'-\lambda_2-c_p(T_1'-T_2)]+\\ +(m'-m'')(s_1'-s_2)+\mu(\lambda_{1e}-q_{e_1}')\end{array}$$

The heat of solution s varies probably with the temperature and the pressure of the gas, but we do not know the law of this variation, and we, therefore, assume this quantity to be constant and equal to 514.3 calories, as found by Favre and Silbermann, for ordinary temperatures and pressures.

Making $s_1' = s_2 = s$ in the above equation it becomes

$$Q' = m[c_p(T_1 - T_1') + s] + (m - m')[\lambda_1' - \lambda_2 - c_p(T_1' - T_2)] + \mu(\lambda^e_1 - q_{e_1}) \quad (115)$$

we have further

$$m = \frac{V_2}{u_2 + \dfrac{0.001}{\delta}} = \frac{V_2}{v_2}.$$

§ 38. The two following tables exhibit results; the two cases of § 36 are taken, supposing that the weight of watery vapor carried over is 5 per cent. of the weight of the gas circulating.

138

First case: $t=-15°$, $m=1^{k}.932$, $m''=0^{k}.3091$, $\mu=0^{k}.0966$.

Temperature of interior of condenser t_1'.	Pressure in condenser P_1.	Temperature in boiler t_1.	Heat carried away from condenser & absorption chamber Q_1+Q_1'.	Disposable negative heat Q_v.	Heat expended Q'.	Ratio of efficie'cy theoretic $\frac{Q}{Q_v}$	Efficie'y per kilogram of coal $4,000\frac{Q}{Q'}$	Weight of gas dissolved by water carried over m'.
Degrees.	Kilog.	Degrees.	Cal.	Cal.	Cal.	Cal.	Cal.	Kilog.
15	74,504	67.77	1,584.43	485.69	1,098.74	0.442	1,768	0.400
20	87,925	81.74	1,583.92	477.47	1,106.45	0.431	1,724	0.407
25	103,073	95.69	1,584.80	470.63	1,114.17	0.422	1,688	0.421
30	120,083	109.61	1,583.59	461.79	1,121.80	0.411	1,644	0.426
35	139,054	123.47	1,577.94	448.70	1,129.24	0.397	1,590	0.412
40	160,112	137.27	1,564.13	427.78	1,136.35	0.376	1,506	0.364

Second case; $t_2 = -30°$, $m = 1^k.023$, $m'' = 0^k.147$, $\mu = 0^k.0511$.

Temperature of interior of condenser t_1.	Temperature in boiler t_1'.	Heat carried away from condenser & absorption chamber $Q_1 + Q_1'$.	Disposable negative heat Q.	Heat expended Q'.	Ratio of efficiency theoretic $\dfrac{Q}{Q'}$	Efficiency per kilogram of coal $4{,}000\dfrac{Q}{Q'}$	Weight of gas dissolved by water carried over m'.
Degrees.	Degrees.	Cal.	Cal.	Cal.	Cal.	Cal.	Kilog.
15	106.07	859.46	258.59	600.87	0.430	1,721	0.212
20	121.61	859.88	254.08	605.80	0.419	1,676	0.215
25	137.13	861.21	250.57	610.64	0.410	1,641	0.223
30	152.61	861.38	245.75	615.63	0.399	1,596	0.225
35	168.03	857.67	238.92	618.75	0.386	1,544	0.218
40	183.38	852.93	227.99	624.94	0.365	1,460	0.193

If we compare the figures of these tables with those of § 36, we find that the cooling of the absorption chamber, and the presence of watery vapor, reduce the efficiency to a considerable extent.

We see further that the useful effect diminishes in proportion as the temperature is lowered in the refrigerant, but that the results remain the same for the same temperature of the condenser.

In the machines employing mechanical power, the efficiency on the other hand diminishes with the temperature of the refrigerant.

§ 39. In the practical manufacture of artificial ice, we estimate the performance at about 1200 or 1500 negative calories for each kilogram of coal burned, which is about 80 per cent. of the above figures. The difference here between theory and practice may fairly be attributed to external losses of temperature, to imperfect action in the exchanges of heat, and to expenditure of work in driving the pumps.

The constructors of sulphur dioxide machines claim a practical result of 2500 calories per horse power per hour. As a good engine consumes two kilograms of coal per horse power per hour, we are afforded a means of comparing the two kinds of apparatus in the matter of economy, and the result is in favor of the chemical action machines. The latter also afford the advantage of low temperatures.

In the sulphur dioxide machine, a lower temperature than $-12°$ is not attained without loss of useful effect, while in the ammonia machine $-25°$ and $-30°$ are readily and economically obtained.

We will not enter here upon questions of a purely practical character which affect the comparative values of the several ice machines, as our object has been simply to establish the theoretic conditions under which they work.

APPENDIX.

NOTE UPON THE DETERMINATION OF THE LATENT HEAT OF VAPORIZATION, ALSO OF THE SPECIFIC HEAT OF SULPHUR DIOXIDE AND AMMONIA IN THE FORM OF LIQUID.

It was shown in section 27 that the relation between the pressure, specific volume and temperature of a liquefiable gas, being represented by the equation

$$Pv = BT - CP^n, \qquad (116)$$

the constants B, C and n can be determined by means of the coefficient of dilatation, and the experiments of Regnault upon the compressibility of gases.

These constants are

	For Sulphur Dioxide.	For Ammonia.
B	13.882	52.4943
C	3.8455	43.7144
n	0.44487	0.32685

Regnault determined also the elastic

forces of these substances at different temperatures, and established the empirical formula

$$\log F = a + b a^t + c \beta^t.$$

This form not being convenient for calculation, we have preferred to take the formula called Roche's

$$P = a \alpha^{\frac{t}{1+mt}} \qquad (117)$$

and we have calculated the three constants a, α and m for both sulphur dioxide and ammonia.

These constants are

	For Sulphur Dioxide.	For Ammonia.
$a =$	15840	43474.64
$\log. a =$	4.1991752	4.6382260
$\alpha =$	1.04135	1.0386605
$\log. \alpha =$	0.0176387	0.0164736
$m =$	0.0043129	0.0040112

Finally M. Regnault found for the specific heat of sulphur dioxide 0.15438, and of ammonia gas 0.50836.

On the other hand Clausius established between the latent heat r, the absolute temperature T, the pressure P, and the quantity u the relation

$$\frac{r}{u} = \mathrm{AT}\frac{d\mathrm{P}}{dt}$$

or
$$\frac{r}{\mathrm{AP}u} = \frac{\mathrm{T}\frac{d\mathrm{P}}{dt}}{\mathrm{P}} \qquad (118)$$

u is the increase of volume of a unit of weight of a volatile liquid when transformed into vapor.

If v is the specific volume of the vapor, we have

$$u = v - \frac{0.001}{\delta},$$

δ being the density of the liquid, and consequently

$$\mathrm{AP}u = \mathrm{ABT} - \mathrm{ACP}^n\frac{0.001.\mathrm{AP}}{\delta} \qquad (119)$$

The constants B, C and n being known, the equation will give APu.

Knowing APu we find r by eq. 118.

$$r = \mathrm{AP}u\frac{\mathrm{T}}{\mathrm{P}}\frac{d\mathrm{P}}{dt}$$

or in consequence of eq. 117

$$r = APu \cdot \frac{T.la}{(1+mt)^2} \qquad (120)$$

The equation 120 will give r in terms of T and APu.

Finally it was shown in § 27 that the quantity λ, that is, the total heat of vaporization satisfies the equation

$$\lambda = \lambda_0 + c_p t - \frac{AC}{n}(P^n - P_0^n)$$

At temperature zero we have

$$\lambda_0 = r_0$$

then it becomes

$$\lambda = r_0 + c_p t - \frac{AC}{n}(P^n - P_0^n) \qquad (121)$$

an equation in which P_0 represents the pressure of the vapor at zero, c_p the specific heat of the vapor at constant pressure, and r_0 the latent heat at zero.

The heat of the liquid

$$q = \lambda - r.$$

We shall have then

$$q = r_0 + c_p t - \frac{AC}{n}(P^n - P_0^n) - APu \frac{Tla}{(1+mt)} \qquad (122)$$

and the specific heat of the liquid

$$c = \frac{dq}{dt}$$

The equations (119), (120), (121) and (122), involving laborious calculations, we can replace the second member by empirical expressions of the form $A' + B't + C't^2$, and then calculate the constants by means of three values taken at the two extremities and middle of the thermometric scale, and previously determined by aid of these equations.

We thus find for

Sulphur Dioxide

$APu = 8,243 + 0,0196t - 0,000116t^2$
$r = 91,396 - 0,2361t - 0,000135t^2$
$\lambda = 91,396 + 0,12723t - 0,000131t^2$
$q = 0,36333t + 0,00004t^2$
$c = 0,36333 + 0,00008t.$

For Ammonia

$APu = 30,154 + 0,08861t - 0,000059t^2$
$r = 313,63 - 0,6250t - 0,002111t^2$
$\lambda = 313,63 + 0,3808t - 0,000282t^2$
$q = 1,0058t + 0,001829t^2$
$c = 1,0058 + 0,003658t.$

The specific heat of liquid ammonia is nearly equal to that of water. This result, though astonishing at first, is comprehended when we reflect that the specific heat of the gas at constant pressure 0.50836 is higher than that of steam (0.4805). It would be interesting to verify by experiment the theoretical conclusion.

The results obtained here for ammonia are, however, only approximate, for we need, in order to determine the constants of eq. (116), the coefficient of dilatation of this gas, and at present it is not known.

To facilitate calculations upon the ice machines, we have prepared the following tables for sulphur dioxide and ammonia. They give for each 5° the heat of the liquid q, the total heat of vaporization λ, the latent heat of vaporization r, the internal latent heat ρ, the external latent heat APu, and the weight of a cubic meter of vapor $\frac{1}{v}$, for the temperatures between $-40°$ and $+40°$.

TABLE I.—SULPHUR DIOXIDE (SATURATED).

Temperature centigrade. t	Absolute temperature. T	Pressure in kilo. per meter. P (Regnault)	Total heat. λ	Heat of liquid. q	Heat of vaporization. r	Latent heat external. APu	Latent heat internal. ρ	u	Weight of vapor per cubic meter. $\frac{1}{v}$
−30	243	3,908	87,461	−10,864	98,325	7,551	90,774	0.822	1,215
−25	348	5,082	88,134	−9,058	97,192	7,681	89,511	0.641	1,558
−20	253	6,519	88,799	−7,251	96,050	7,805	88,245	0.507	1,967
−15	258	8,265	89,458	−5,441	94,899	7,923	86,976	0.406	2,456
−10	263	10,366	90,111	−3,629	93,740	8,035	85,705	0.328	3,038
−5	268	12,874	90,757	−1,814	92,571	8,142	84,429	0.268	3,719
0	273	15,840	91,376	0,000	91,396	8,243	83,153	0.221	4,518
5	278	19,322	92,029	1,818	90,211	8,338	81,873	0.183	5,447
10	283	23,378	92,655	3,637	89,018	8,427	80,591	0.153	6,514
15	288	28,074	93,275	5,459	87,816	8,511	79,305	0.129	7,740
20	293	33,474	93,888	7,283	86,605	8,589	78,016	0.109	9,132
25	298	39,645	94,495	9,108	85,387	8,661	76,726	0.093	10,718
30	303	46,659	95,095	10,936	84,159	8,727	75,432	0.079	12,500
35	308	54,585	95,689	12,765	82,924	8,787	74,137	0.068	14,513
40	—	63,496	96,276	14,597	81,679	8,841	72,838	0.057	16,750

TABLE II.—AMMONIA GAS SATURATED.

Temperature centigrade. t	Absolute temperature. T	Pressure in kilo. per meter. P (Regnault)	Total heat. λ	Heat of liquid. q	Heat of vaporization. r	Latent heat external. APu	Latent heat internal. ρ	u	Weight of vapor per cubic meter. $\frac{1}{v}$
—40	233	7,187	297,967	—37,268	335,235	26,515	308,720	1.564	0.639
—35	238	9,302	299,955	—32,963	332,918	26,980	305,938	1.230	0.812
—30	243	11,918	301,952	—28,528	330,480	27,438	303,042	0.976	1.023
—25	248	15,120	303,934	—23,002	327,936	27,902	300,034	0.782	1.276
—20	253	19,003	305,901	—19,384	325,285	28,358	296,927	0.633	1.577
—15	258	23,669	307,855	—14,676	322,531	28,812	293,719	0.516	1.932
—10	263	29,225	309,794	—9,875	319,669	29,262	290,407	0.424	2.348
—5	268	35,797	311,719	—4,983	316,702	29,708	286,994	0.352	2.840
0	273	43,475	313,630	0,000	313,630	30,154	283,476	0.294	3.385
5	278	52,405	315,527	5,075	310,452	30,594	279,858	0.247	4.020
10	283	62,707	317,418	10,241	307,177	31,034	276,143	0.210	4.737
15	288	74,504	319,279	15,498	303,781	31,470	272,311	0.179	5.543
20	293	87,925	321,133	20,848	300,285	31,902	268,383	0.154	6.447
25	298	103,073	322,979	26,288	296,691	32,332	264,359	0.133	7.446
30	303	120,083	324,800	31,820	292,980	32,754	260,226	0.116	8.547
35	308	139,054	326,611	37,443	289,168	33,182	255,986	0.101	9.766
40	313	160,112	328,431	43,196	285,235	33,604	251,631	0.089	11.074

₊ *Any book in this Catalogue sent free by mail, on receipt of price.*

VALUABLE
SCIENTIFIC BOOKS,
PUBLISHED BY
D. VAN NOSTRAND,
23 Murray Street, and 27 Warren Street,

NEW YORK.

WEISBACH. A MANUAL OF THEORETICAL MECHANICS. By Julius Weisbach, Ph. D. Translated by Eckley B. Coxe, A.M., M.E. 1100 pages and 902 wood-cut illustrations. 8vo, cloth, ... $10 00

FRANCIS. LOWELL HYDRAULIC EXPERIMENTS—being a Selection from Experiments on Hydraulic Motors, on the Flow of Water over Weirs, and in open Canals of Uniform Rectangular Section, made at Lowell, Mass. By J. B. Francis, Civil Engineer Third edition, revised and enlarged, with 23 copper-plates, beautifully engraved, and about 100 new pages of text. 4to, cloth, ... 15 00

KIRKWOOD ON THE FILTRATION OF RIVER WATERS, for the Supply of Cities, as practised in Europe. By James P. Kirkwood. Illustrated by 30 double-plate engravings. 4to, cloth, ... 15 00

1

D. VAN NOSTRAND'S PUBLICATIONS.

FANNING. A Practical Treatise of Water Supply Engineering. Relating to the Hydrology, Hydrodynamics, and Practical Construction of Water-Works, in North America. With numerous Tables and 180 illustrations. By J. T. Fanning, C.E. 650 pages. 8vo, cloth extra, . $6 00

WHIPPLE. An Elementary Treatise on Bridge Building. By S. Whipple, C. E. New Edition. Illustrated. 8vo, cloth, 4 00

MERRILL. Iron Truss Bridges for Railroads. The Method of Calculating Strains in Trusses, with a careful comparison of the most prominent Trusses, in reference to economy in combination, etc., etc. By Bvt. Col. William E. Merrill, U. S. A., Corps of Engineers. Nine lithographed plates of illustrations. Third edition. 4to, cloth, ... 5 00

SHREVE. A Treatise on the Strength of Bridges and Roofs. Comprising the determination of Algebraic formulas for Strains in Horizontal, Inclined or Rafter, Triangular, Bowstring Lenticular and other Trusses, from fixed and moving loads, with practical applications and examples, for the use of Students and Engineers By Samuel H. Shreve, A. M., Civil Engineer. Second edition, 87 woodcut illustrations. 8vo, cloth, . . . 5 00

KANSAS CITY BRIDGE. With an Account of the Regimen of the Missouri River,— and a description of the Methods used for Founding in that River. By O. Chanute, Chief Engineer, and George Morison, Assistant Engineer. Illustrated with five lithographic views and twelve plates of plans. 4to, cloth, ... 6 00

D. VAN NOSTRAND'S PUBLICATIONS.

CLARKE. DESCRIPTION OF THE IRON RAILWAY BRIDGE Across the Mississippi River at Quincy, Illinois. By Thomas Curtis Clarke, Chief Engineer. With twenty-one lithographed Plans. 4to, cloth, ... $7 50

ROEBLING. LONG AND SHORT SPAN RAILWAY BRIDGES. By John A. Roebling, C. E. With large copperplate engravings of plans and views. Imperial folio, cloth, . 25 00

DUBOIS. THE NEW METHOD OF GRAPHICAL STATICS. By A. J. Dubois, C. E., Ph. D. 60 illustrations. 8vo, cloth, ... 1 50

McELROY. PAPERS ON HYDRAULIC ENGINEERING. The Hempstead Storage Reservoir of Brooklyn, its Engineering Theory and Results, By Samuel McElroy, C. E. 8vo, paper, . 50

BOW. A TREATISE ON BRACING—with its application to Bridges and other Structures of Wood or Iron. By Robert Henry Bow, C. E. 156 illustrations on stone. 8vo, cloth, 1 50

STONEY. THE THEORY OF STRAINS IN GIRDERS—and Similar Structures—with Observations on the Application of Theory to Practice, and Tables of Strength and other Properties of Materials. By Bindon B. Stoney, B. A. New and Revised Edition, with numerous illustrations. Royal 8vo, 664 pp., cloth, ... 12 50

HENRICI. SKELETON STRUCTURES, especially in their Application to the building of Steel and Iron Bridges. By Olaus Henrici. 8vo, cloth, ... 1 50

KING. LESSONS AND PRACTICAL NOTES ON STEAM. The Steam Engine, Propellers, &c., &c., for Young Engineers. By the late W. R. King, U. S. N., revised by Chief-Engineer J. W. King, U. S. Navy. 19th edition. 8vo, cloth, ... 2 00

D. VAN NOSTRAND'S PUBLICATIONS.

AUCHINCLOSS. Application of the Slide Valve and Link Motion to Stationary, Portable, Locomotive and Marine Engines. By William S Auchincloss. Designed as a hand-book for Mechanical Engineers. With 37 wood-cuts and 21 lithographic plates, with copper-plate engraving of the Travel Scale. Sixth edition. 8vo, cloth,   **$3 00**

BURGH. Modern Marine Engineering, applied to Paddle and Screw Propulsion. Consisting of 36 Colored Plates, 259 Practical Wood-cut Illustrations, and 403 pages of Descriptive Matter, the whole being an exposition of the present practice of the following firms: Messrs J. Penn & Sons; Messrs. Maudslay, Sons & Field; Messrs. James Watt & Co.; Messrs. J. & G. Rennie. Messrs. R. Napier & Sons; Messrs J. & W. Dudgeon; Messrs. Ravenhill & Hodgson; Messrs Humphreys & Tenant; Mr J. T. Spencer, and Messrs. Forrester & Co. By N P. Burgh, Engineer. One thick 4to vol., cloth, $25.00; half morocco, **30 00**

BACON. A Treatise on the Richard's Steam-Engine Indicator — with directions for its use. By Charles T. Porter. Revised, with notes and large additions as developed by American Practice; with an Appendix containing useful formulæ and rules for Engineers. By F. W. Bacon, M. E. Illustrated Second edition. 12mo. Cloth $1.00; morocco,   **1 50**

ISHERWOOD Engineering Precedents for Steam Machinery. By B. F. Isherwood, Chief Engineer, U. S. Navy. With illustrations. Two vols. in one. 8vo, cloth,   **2 50**

STILLMAN. The Steam Engine Indicator — and the Improved Manometer Steam and Vacuum Gauges—their utility and application. By Paul Stillman. New edition. 12mo, cloth,   **1 00**

D. VAN NOSTRAND'S PUBLICATIONS.

MacCORD. A PRACTICAL TREATISE ON THE SLIDE VALVE, BY ECCENTRICS—examining by methods the action of the Eccentric upon the Slide Valve, and explaining the practical processes of laying out the movements, adapting the valve for its various duties in the steam-engine. By C. W Mac Cord, A. M., Professor of Mechanical Drawing, Stevens' Institute of Technology, Hoboken, N. J. Illustrated. 4to, cloth. $3 00

PORTER. A TREATISE ON THE RICHARDS' STEAM-ENGINE INDICATOR, and the Development and Application of Force in the Steam-Engine. By Charles T. Porter. Third edition, revised and enlarged. Illustrated. 8vo, cloth, . . . 3 50

McCULLOCH A TREATISE ON THE MECHANICAL THEORY OF HEAT, AND ITS APPLICATIONS TO THE STEAM-ENGINE. By Prof R. S. McCulloch, of the Washington and Lee University, Lexington. Va. 8vo, cloth, ... 3 50

VAN BUREN. INVESTIGATIONS OF FORMULAS—for the Strength of the Iron parts of Steam Machinery. By J D. Van Buren, Jr., C. E. Illustrated. 8vo, cloth, . 2 00

STUART. HOW TO BECOME A SUCCESSFUL ENGINEER. Being Hints to Youths intending to adopt the Profession. By Bernard Stuart, Engineer. Sixth edition 18mo, boards, ... 50

SHIELDS. NOTES ON ENGINEERING CONSTRUCTION. Embracing Discussions of the Principles involved, and Descriptions of the Material employed in Tunneling, Bridging, Canal and Road Building, etc., etc. By J. E. Shields, C. E. 12mo. cloth, . . . 1 50

D. VAN NOSTRAND'S PUBLICATIONS.

WEYRAUCH. STRENGTH AND CALCULATION OF DIMENSIONS OF IRON AND STEEL CONSTRUCTIONS. Translated from the German of J. J. Weyrauch, Ph. D., with four folding Plates. 12mo, cloth, . . . $1 00

STUART. THE NAVAL DRY DOCKS OF THE UNITED STATES. By Charles B. Stuart, Engineer in Chief, U. S. Navy. Twenty-four engravings on steel. Fourth edition. 4to, cloth, ... 6 00

COLLINS. THE PRIVATE BOOK OF USEFUL ALLOYS, and Memoranda for Goldsmiths, Jewellers, etc. By James E. Collins. 18mo, flexible cloth, ... 50

TUNNER. A TREATISE ON ROLL-TURNING FOR THE MANUFACTURE OF IRON. By Peter Tunner Translated by John B. Pearse. With numerous wood-cuts, 8vo, and a folio Atlas of 10 lithographed plates of Rolls, Measurements, &c. Cloth, . . 10 00

GRÜNER. THE MANUFACTURE OF STEEL. By M. L. Gruner. Translated from the French, by Lenox Smith, A.M., E.M.; with an Appendix on the Bessemer Process in the United States, by the translator. Illustrated by lithographed drawings and wood-cuts. 8vo, cloth, ... 3 50

BARBA. THE USE OF STEEL IN CONSTRUCTION. Methods of Working, Applying, and Testing Plates and Bars. By J. Barba. Translated from the French, with a Preface by A. L. Holley, P.B. Illustrated. 12mo, cloth, 1 50

BELL. CHEMICAL PHENOMENA OF IRON SMELTING. An Experimental and Practical Examination of the Circumstances which Determine the Capacity of the Blast Furnace, the Temperature of the Air, and the Proper Condition of the Materials to be operated upon. By I. Lowthian Bell. 8vo, cloth, ... 6 00

D. VAN NOSTRAND'S PUBLICATIONS.

WARD. STEAM FOR THE MILLION. A Popular Treatise on Steam and its Application to the Useful Arts, especially to Navigation. By J. H. Ward, Commander U. S. Navy. 8vo, cloth, ... $1 00

CLARK. A MANUAL OF RULES, TABLES AND DATA FOR MECHANICAL ENGINEERS. Based on the most recent investigations. By Dan. Kinnear Clark. Illustrated with numerous diagrams. 1012 pages. 8vo. Cloth, $7 50; half morocco, ... 10 00

JOYNSON. THE METALS USED IN CONSTRUCTION: Iron, Steel, Bessemer Metals, etc, By F. H. Joynson. Illustrated. 12mo, cloth, ... 75

DODD. DICTIONARY OF MANUFACTURES, MINING, MACHINERY, AND THE INDUSTRIAL ARTS. By George Dodd. 12mo, cloth, 1 50

VON COTTA. TREATISE ON ORE DEPOSITS. By Bernhard Von Cotta, Freiburg, Saxony. Translated from the second German ed., by Frederick Prime, Jr., and revised by the author. With numerous illustrations. 8vo, cloth, ... 4 00

PLATTNER. MANUAL OF QUALITATIVE AND QUANTITATIVE ANALYSIS WITH THE BLOWPIPE. From the last German edition. Revised and enlarged. By Prof. Th. Richter, o the Royal Saxon Mining Academy. Translated by Professor H. B. Cornwall. With eighty-seven wood-cuts and lithographic plate. Third edition, revised. 568 pp. 8vo, cloth, ... 5 00

PLYMPTON. THE BLOW-PIPE: A Guide to its Use in the Determination of Salts and Minerals. Compiled from various sources, by George W. Plympton, C. E., A. M., Professor of Physical Science in the Polytechnic Institute, Brooklyn, N. Y. 12mo, cloth, 1 50

D. VAN NOSTRAND'S PUBLICATIONS.

JANNETTAZ. A GUIDE TO THE DETERMINATION OF ROCKS; being an Introduction to Lithology. By Edward Jannettaz, Docteur des Sciences. Translated from the French by G. W. Plympton, Professor of Physical Science at Brooklyn Polytechnic Institute. 12mo, cloth, ... $1 50

MOTT. A PRACTICAL TREATISE ON CHEMISTRY (Qualitative and Quantitative Analysis), Stoichiometry, Blowpipe Analysis, Mineralogy, Assaying, Pharmaceutical Preparations. Human Secretions, Specific Gravities, Weights and Measures, etc., etc., etc. By Henry A. Mott, Jr., E. M., Ph. D. 650 pp. 8vo, cloth, ... 6 00

PYNCHON. INTRODUCTION TO CHEMICAL PHYSICS; Designed for the Use of Academies, Colleges, and High Schools. Illustrated with numerous engravings, and containing copious experiments, with directions for preparing them. By Thomas Ruggles Pynchon, D. D., M. A., President of Trinity College, Hartford. New edition, revised and enlarged. Crown 8vo, cloth, . . . 3 00

PRESCOTT. CHEMICAL EXAMINATION OF ALCOHOLIC LIQUORS. A Manual of the Constituents of the Distilled Spirits and Fermented Liquors of Commerce, and their Qualitative and Quantitative Determinations. By Alb. B. Prescott, Prof. of Chemistry, University of Michigan. 12mo, cloth, . 1 50

ELIOT AND STORER. A COMPENDIOUS MANUAL OF QUALITATIVE CHEMICAL ANALYSIS. By Charles W. Eliot and Frank H. Storer. Revised, with the co-operation of the Authors, by William Ripley Nichols, Professor of Chemistry in the Massachusetts Institute of Technology. New edition, revised. Illustrated. 12mo, cloth, ... 1 50

D. VAN NOSTRAND'S PUBLICATIONS.

NAQUET. LEGAL CHEMISTRY. A Guide to the Detection of Poisons, Falsification of Writings, Adulteration of Alimentary and Pharmaceutical Substances; Analysis of Ashes, and Examination of Hair, Coins, Fire-arms and Stains, as Applied to Chemical Jurisprudence. For the Use of Chemists, Physicians, Lawyers, Pharmacists, and Experts. Translated, with additions, including a List of Books and Memoirs on Toxicology, etc., from the French of A. Naquet, by J. P. Battershall, Ph. D.; with a Preface by C. F. Chandler, Ph. D., M. D., LL. D. Illustrated. 12mo, cloth, ... $2 00

PRESCOTT. OUTLINES OF PROXIMATE ORGANIC ANALYSIS for the Identification, Separation, and Quantitative Determination of the more commonly occurring Organic Compounds. By Albert B. Prescott, Professor of Chemistry, University of Michigan. 12mo, cloth, ... 1 75

DOUGLAS AND PRESCOTT. QUALITATIVE CHEMICAL ANALYSIS. A Guide in the Practical Study of Chemistry, and in the work of Analysis. By S. H. Douglas and A. B. Prescott; Professors in the University of Michigan. Second edition, revised. 8vo, cloth, ... 3 50

RAMMELSBERG. GUIDE TO A COURSE OF QUANTITATIVE CHEMICAL ANALYSIS, ESPECIALLY OF MINERALS AND FURNACE PRODUCTS. Illustrated by Examples. By C. F. Rammelsberg. Translated by J. Towler, M. D. 8vo, cloth, ... 2 25

BEILSTEIN. AN INTRODUCTION TO QUALITATIVE CHEMICAL ANALYSIS. By F. Beilstein. Third edition. Translated by I. J. Osbun. 12mo. cloth, ... 75

POPE. A Hand-book for Electricians and Operators. By Frank L. Pope. Ninth edition. Revised and enlarged, and fully illustrated. 8vo, cloth, ... 2 00

D. VAN NOSTRAND'S PUBLICATIONS.

SABINE. HISTORY AND PROGRESS OF THE ELECTRIC TELEGRAPH, with Descriptions of some of the Apparatus. By Robert Sabine, C. E. Second edition. 12mo, cloth, . . $1 25

DAVIS AND RAE. HAND BOOK OF ELECTRICAL DIAGRAMS AND CONNECTIONS. By Charles H. Davis and Frank B. Rae. Illustrated with 32 full-page illustrations. Second edition. Oblong 8vo, cloth extra, . . . 2 00

HASKINS. THE GALVANOMETER, AND ITS USES. A Manual for Electricians and Students. By C. H. Haskins. Illustrated. Pocket form, morocco, ... 1 50

LARRABEE. CIPHER AND SECRET LETTER AND TELEGRAPAIC CODE, with Hogg's Improvements. By C. S. Larrabee. 18mo, flexible cloth, ... 1 00

GILLMORE PRACTICAL TREATISE ON LIMES, HYDRAULIC CEMENT. AND MORTARS. By Q. A. Gillmore, Lt.-Col. U. S. Engineers, Brevet Major-General U. S. Army. Fifth edition, revised and enlarged. 8vo, cloth, 4 00

GILLMORE. COIGNET BETON AND OTHER ARTIFICIAL STONE. By Q. A. Gillmore, Lt. Col. U. S. Engineers, Brevet Major-General U. S. Army. Nine plates, views, etc. 8vo, cloth, ... 2 50

GILLMORE. A PRACTICAL TREATISE ON THE CONSTRUCTION OF ROADS, STREETS, AND PAVEMENTS. By Q. A. Gillmore, Lt.-Col. U. S. Engineers, Brevet Major-General U. S. Army. Seventy illustrations. 12mo, clo., 2 00

GILLMORE. REPORT ON STRENGTH OF THE BUILDING STONES IN THE UNITED STATES, etc. 8vo, cloth, ... 1 00

HOLLEY. AMERICAN AND EUROPEAN RAILWAY PRACTICE, in the Economical Generation of Steam. By Alexander L. Holley. B. P. With 77 lithographed plates. Folio, cloth, 12 00

D. VAN NOSTRAND'S PUBLICATIONS.

HAMILTON. USEFUL INFORMATION FOR RAILWAY MEN. Compiled by W. G. Hamilton, Engineer. Seventh edition, revised and enlarged. 577 pages. Pocket form, morocco, gilt, ... , . $2 00

STUART. THE CIVIL AND MILITARY ENGINEERS OF AMERICA. By General Charles B. Stuart, Author of "Naval Dry Docks of the United States," etc., etc. With nine finely-executed Portraits on steel, of eminent Engineers, and illustrated by Engravings of some of the most important and original works constructed in America. 8vo, cloth, ... 5 00

ERNST. A MANUAL OF PRACTICAL MILITARY ENGINEERING. Prepared for the use of the Cadets of the U. S. Military Academy, and for Engineer Troops. By Capt. O. H. Ernst, Corps of Engineers, Instructor in Practical Military Engineering, U. S. Military Academy. 193 wood-cuts and 3 lithographed plates. 12mo, cloth, . . 5 00

SIMMS. A TREATISE ON THE PRINCIPLES AND PRACTICE OF LEVELLING, showing its application to purposes of Railway Engineering and the Construction of Roads, etc. By Frederick W. Simms, C. E. From the fifth London edition, revised and corrected, with the addition of Mr. Law's Practical Examples for Setting-out Railway Curves. Illustrated with three lithographic plates, and numerous wood-cuts. 8vo, cloth, ... 2 50

JEFFERS. NAUTICAL SURVEYING. By William N. Jeffers, Captain U. S. Navy. Illustrated with 9 copperplates, and 31 wood-cut illustrations. 8vo, cloth, ... 5 00

THE PLANE TABLE. ITS USES IN TOPOGRAPHICAL SURVEYING. From the papers of the U. S. Coast Survey. 8vo, cloth, . . 2 00

D. VAN NOSTRAND'S PUBLICATIONS.

A TEXT-BOOK ON SURVEYING, Projections, and Portable Instruments, for the use of the Cadet Midshipmen, at the U. S. Naval Academy. 9 lithographed plates, and several wood-cuts. 8vo, cloth, . . . $2 00

CHAUVENET. New Method of Correcting Lunar Distances. By Wm. Chauvenet, LL.D. 8vo, cloth, ... 2 00

BURT. Key to the Solar Compass, and Surveyor's Companion; comprising all the Rules necessary for use in the Field. By W. A. Burt, U. S. Deputy Surveyor. Second edition. Pocket-book form, tuck, . 2 50

HOWARD. Earthwork Mensuration on the Basis of the Prismoidal Formulæ. Containing simple and labor-saving method of obtaining Prismoidal Contents directly from End Areas. Illustrated by Examples, and accompanied by Plain Rules for practical uses. By Conway R. Howard, Civil Engineer, Richmond, Va. Illustrated. 8vo, cloth, ... 1 50

MORRIS. Easy Rules for the Measurement of Earthworks, by means of the Prismoidal Formulæ. By Elwood Morris, Civil Engineer. 78 illustrations. 8vo, cloth, 1 50

CLEVENGER. A Treatise on the Method of Government Surveying, as prescribed by the U. S. Congress and Commissioner of the General Land Office. With complete Mathematical, Astronomical, and Practical Instructions for the use of the U. S. Surveyors in the Field. By S. V. Clevenger, U. S. Deputy Surveyor. Illustrated. Pocket form, morocco, gilt, . . . 2 50

HEWSON. Principles and Practice of Embanking Lands from River Floods, as applied to the Levees of the Mississipi. By William Hewson, Civil Engineer. 8vo, cloth, ... 2 00

D. VAN NOSTRAND'S PUBLICATIONS.

MINIFIE. A TEXT-BOOK OF GEOMETRICAL DRAWING, for the use of Mechanics and Schools. With Illustrations for Drawing Plans, Elevations of Buildings and Machinery. With over 200 diagrams on steel. By William Minifie, Architect. Ninth edition. Royal 8vo, cloth, ... $4 00

MINIFIE. GEOMETRICAL DRAWING. Abridged from the octavo edition, for the use of Schools. Illustrated with 48 steel plates. New edition, enlarged. 12mo, cloth, 2 00

FREE HAND DRAWING. A GUIDE TO ORNAMENTAL, Figure, and Landscape Drawing. By an Art Student. Profusely illustrated. 18mo, boards, ... 50

AXON. THE MECHANIC'S FRIEND. A Collection of Receipts and Practical Suggestions, relating to Aquaria—Bronzing—Cements—Drawing—Dyes—Electricity—Gilding—Glass-working—Glues—Horology—Lacquers—Locomotives—Magnetism—Metal-working—Modelling—Photography—Pyrotechny—Railways—Solders—Steam-Engine—Telegraphy—Taxidermy—Varnishes—Waterproofing-and Miscellaneous Tools, Instruments, Machines, and Processes connected with the Chemical and Mechanical Arts. By William E. Axon, M.R.S.L. 12mo, cloth. 300 illustrations, . . . 1 50

HARRISON. MECHANICS' TOOL BOOK, with Practical Rules and Suggestions, for the use of Machinists, Iron Workers, and others. By W. B. Harrison. 44 illustrations. 12mo, cloth 1 50

JOYNSON. THE MECHANIC'S AND STUDENT'S GUIDE in the designing and Construction of General Machine Gearing. Edited by Francis H. Joynson. With 18 folded plates. 8vo, cloth 2 00

D. VAN NOSTRAND'S PUBLICATIONS.

RANDALL. QUARTZ OPERATOR'S HAND-BOOK. By P. M. Randall. New Edition. Revised and Enlarged. Fully illustrated. 12mo, cloth, ... $2 00

LORING. A HAND-BOOK ON THE ELECTRO-MAGNETIC TELEGRAPH. By A. E. Loring. 18mo, illustrated. Paper boards, 50 cents; cloth, 75 cents; morocco, ... 1 00

BARNES. SUBMARINE WARFARE, DEFENSIVE AND OFFENSIVE. Descriptions of the various forms of Torpedoes, Submarine Batteries and Torpedo Boats actually used in War. Methods of Ignition by Machinery, Contact Fuzes, and Electricity, and a full account of experiments made to determine the Explosive Force of Gunpowder under Water. Also a discussion of the Offensive Torpedo system; its effect upon Iron-clad Ship systems, and influence upon future Naval Wars. By Lieut.-Com. John S. Barnes, U. S. N. With 20 lithographic plates and many wood-cuts. 8vo, cloth, 5 00

FOSTER. SUBMARINE BLASTING, in Boston Harbor, Mass. Removal of Tower and Corwin Rocks. By John G. Foster, U. S. Eng. and Bvt. Major General U. S. Army. With seven Plates. 4to, cloth, 3 50

PLYMPTON. THE ANEROID BAROMETER: Its Construction and Use, compiled from several sources. 16mo, boards, illustrated, 50 cents; morocco, ... 1 00

WILLIAMSON. ON THE USE OF THE BAROMETER ON SURVEYS AND RECONNAISSANCES. Part I.–Meteorology in its Connection with Hypsometry. Part II.–Barometric Hypsometry. By R. S. Williamson, Bvt. Lt.-Col. U.S.A., Major Corps of Engineers. With illustrative tables and engravings. 4to, cloth, ... 15 00

D. VAN NOSTRAND'S PUBLICATIONS.

WILLIAMSON. PRACTICAL TABLES IN METE-OROLOGY AND HYPSOMETRY, in connection with the use of the Barometer By Col. R. S. Williamson, U. S. A. 4to, flexible cloth, $2 50

BUTLER. PROJECTILES AND RIFLED CANNON A Critical Discussion of the Principal Systems of Rifling and Projectiles, with Practical Suggestions for their Improvement. By Capt. John S. Butler, Ordnance Corps, U. S. A. 36 Plates. 4to, cloth, . . . 7 50

BENET. ELECTRO-BALLISTIC MACHINES, and the Schultz Chronoscope. By Lt.-Col S. V. Benet, Chief of Ordnance U. S. A. Second edition, illustrated. 4to, cloth, . 3 00

MICHAELIS. THE LE BOULENGE CHRONOGRAPH. With three lithographed folding plates of illustrations. By Bvt. Captian O. E. Michaelis, Ordnance Corpse, U. S. A. 4to, cloth, ... 3 00

NUGENT. TREATISE ON OPTICS; or Light and Sight, theoretically and practically treated; with the application to Fine Art and Industrial Pursuits. By E. Nugent. With 103 illustrations. 12mo, cloth, . . . 1 50

PEIRCE. SYSTEM OF ANALYTIC MECHANICS. By Benjamin Peirce, Professor of Astronomy and Mathematics in Harvard University. 4to. cloth, ... 10 00

CRAIG. WEIGHTS AND MEASURES. An Account of the Decimal System, with Tables of Conversion for Commercial and Scientific Uses. By B. F. Craig, M. D. Square 32mo, limp cloth, ... , . . 50

ALEXANDER. UNIVERSAL DICTIONARY OF WEIGHTS AND MEASURES, Ancient and Modern, reduced to the standards of the United States of America. By J. H. Alexander. New edition. 8vo, cloth, . . 3 50

D. VAN NOSTRAND'S PUBLICATIONS.

ELLIOT. EUROPEAN LIGHT-HOUSE SYSTEMS. Being a Report of a Tour of Inspection made in 1873. By Major George H. Elliot, U. S. Engineers. 51 engravings and 21 wood-cuts. 8vo, cloth, ... $5 00

SWEET. SPECIAL REPORT ON COAL. By S. H. Sweet. With Maps. 8vo, cloth, . . 3 00

COLBURN. GAS WORKS OF LONDON. By Zerah Colburn. 12mo, boards, ... 60

WALKER. NOTES ON SCREW PROPULSION, its Rise and History. By Capt. W. H. Walker, U. S. Navy. 8vo, cloth, . . . 75

POOK. METHOD OF PREPARING THE LINES AND DRAUGHTING VESSELS PROPELLED BY SAIL OR STEAM, including a Chapter on Laying-off on the Mould-loft Floor. By Samuel M. Pook, Naval Constructor. Illustrated. 8vo, cloth, ... 5 00

SAELTZER. TREATISE ON ACOUSTICS in connection with Ventilation. By Alexander Saeltzer. 12mo, cloth, ... 2 00

EASSIE. A HAND-BOOK FOR THE USE OF CONTACTORS, Builders, Architects, Engineers, Timber Merchants, etc., with information for drawing up Designs and Estimates. 250 illustrations. 8vo, cloth, . . . 1 50

SCHUMANN. A MANUAL OF HEATING AND VENTILATION IN ITS PRACTICAL APPLICATION for the use of Engineers and Architects, embracing a series of Tables and Formulæ for dimensions of heating, flow and return Pipes for steam and hot water boilers, flues, etc. etc. By F Schumann, C. E., U. S. Treasury Department 12mo. Illustrated. Full roan, ... 1 50

TONER. DICTIONARY OF ELEVATIONS AND CLIMATIC REGISTER OF THE UNITED STATES. By J. M. Toner, M D. 8vo. Paper, $3.00; cloth. 3 75

D. VAN NOSTRAND'S NEW PUBLICATIONS.

PLYMPTON. THE STAR FINDER OR PLANISPHERE, WITH MOVABLE HORIZON. Arranged by Prof. G. W. Plympton, A.M. Printed in colors on fine card-board, and in accordance with Proctor's Star Atlas, ... **$1 00**

CALDWELL & BRENEMAN. MANUAL OF INTRODUCTORY CHEMICAL PRACTICE, for the use of Students in Colleges and Normal and High Schools. By Prof. George C. Caldwell, and A. A. Breneman, of Cornell University. Second edition, revised and corrected. 8vo, cloth, illustrated. New and enlarged edition, . . **1 50**

SCOFFERN, TRURAN, Etc. THE USEFUL METALS AND THEIR ALLOYS, employed in the conversion of Iron, Copper, Tin, Zinc, Antimony, and Lead ores, with their applications to the Industrial Arts. By John Scoffern, William Truran, etc. Fifth edition, 8vo, half-calf, . . **3 75**

ROSE. THE PATTERN MAKER'S ASSISTANT, embracing Lathe Work, Branch Work, Core Work, Sweep Work, and Practical Gear Constructions, the Preparation and Use of Tools, together with a large collection of useful and valuable Tables. By Joshua Rose, M.E. Illustrated with 250 engravings. 8vo, cloth, . . . **2 50**

SCRIBNER. ENGINEERS' AND MECHANICS' COMPANION, comprising United States Weights and Measures; Mensuration of Superfices and Solids; Tables of Squares and Cubes; Square and Cube Roots; Circumference and Areas of Circles; the Mechanical Powers; Centers of Gravity; Gravitation of Bodies; Pendulums; Specific Gravity of Bodies; Strength, Weight, and Crush of Materials; Water Wheels; Hydrostatics; Hydraulics; Statics; Centers of Percussion and Gyration; Friction Heat; Tables of the Weight of Metals; Scantling, etc.; Steam and the Steam Engine. By J. M. Scribner, A.M. 18th ed. revised, 16mo, full morocco, **1 50**

D. VAN NOSTRAND'S NEW PUBLICATIONS.

SCRIBNER. ENGINEERS', CONTRACTORS' AND SURVEYORS' POCKET TABLE-BOOK: Comprising Logarithms of Numbers, Logarithmic Signs and Tangents, Natural Signs and Natural Tangents, the Traverse Table, and a full and complete set of Excavation and Embankment Tables, together with numerous other valuable tables for Engineers, etc. By T. M. Scribner, A. M. 10th ed. revised, 16mo, full morocco, **$1 50**

EDDY. RESEARCHES IN GRAPHICAL STATICS, embracing New Constructions in Graphical Statics, a new General Method in Graphical Statics, and the Theory of Internal Stress in Graphical Statics. By Prof. Henry T. Eddy, of the University of Cincinnati. 8vo, cloth, . **1 50**

HUXLEY, BARKER, Etc. HALF HOURS WITH MODERN SCIENTISTS. Lectures and Essays. By Professors Huxley, Barker, Stirling, Cope, Tyndall, Wallace, Roscoe, Huggins, Lockyer, Young, Mayer, and Reed. Being the University Series bound up. With a general introduction by Noah Porter, President of Yale College. 2 vols. 12mo, cloth, illustrated. . **2 50**

SHUNK. THE FIELD ENGINEER. A handy book of Practice in the Survey, Location, and Trackwork of Railroads, containing a large collection of Rules and Tables, original and selected, applicable to both the Standard and Narrow Gauge, and prepared with special reference to the wants of the Young Engineer. By Wm. Findlay Shunk, C. E., Chief Engineer of the Construction of the Metropolitan Elevated Railroad. (In Press.) 12mo, morocco.

ADAMS. SEWERS AND DRAINS FOR POPULOUS DISTRICTS. Embracing Rules and Formulas for the dimensions and construction of works of Sanitary Engineers. By Julius Adams, C. E. (In Press.) 8vo, cloth.

D. VAN NOSTRAND'S PUBLICATIONS.

WANKLYN. MILK ANALYSIS. A Practical Treatise on the Examination of Milk, and its Derivatives, Cream, Butter, and Cheese. By J. Alfred Wanklyn, M.R.C.S. 12mo, cloth, ... $1 00

RICE & JOHNSON. ON A NEW METHOD OF OBTAINING THE DIFFERENTIALS OF FUNCTIONS, with especial reference to the Newtonian Conception of Rates or Velocities. By J. Minot Rice, Prof. of Mathematics, U. S Navy, and W. Woolsey Johnson, Prof. of Mathemathics, St. John's College, Annapolis. 12mo, paper. 50

COFFIN. NAVIGATION AND NAUTICAL ASTRONOMY. Prepared for the use of the U. S. Naval Academy. By J. H. C. Coffin, Professor of Astronomy, Navigation and Surveying; with 52 wood-cut illustrations. Fifth edition. 12mo, cloth. . . 3 50

CLARK. THEORETICAL NAVIGATION AND NAUTICAL ASTRONOMY. By Lewis Clark, Lieut.-Commander, U S Navy Illustrated with 41 wood-cuts, including the Vernier. 8vo. cloth, . . 3 00

ROGERS. THE GEOLOGY OF PENNSYLVANIA. By Henry Darwin Rogers, late State Geologist of Pennsylvania. 3 vols 4to, with Portfolio of Maps. Cloth, . . . 30 00

IN PREPARATION.

WEISBACH. MECHANICS OF ENGINEERING, APPLIED MECHANICS; containing Arches, Bridges, Foundations, Hydraulics, Steam Engine, and other Prime Movers, &c., &c. Translated from the latest German Edition 2 vols, 8vo.

Van Nostrand's Science Series.

It is the intention of the Publisher of this Series to issue them at intervals of about a month. They will be put up in a uniform, neat, and attractive form, 18mo, fancy boards. The subjects will be of an eminently scientific nature, and embrace as wide a range of topics as possible,—all of the highest character.

Price, 50 Cents Each.

I. CHIMNEYS FOR, FURNACES, FIRE-PLACES, AND STEAM BOILERS. By R. Armstrong, C. E.

II. STEAM BOILER EXPLOSIONS. By Zerah Colburn.

III. PRACTICAL DESIGNING OF RETAINING WALLS. By Arthur Jacob, A. B. Illustrated.

IV. PROPORTIONS OF PINS USED IN BRIDGES. By Charles E. Bender, C. E. Illustrated.

V. VENTILATION OF BUILDINGS. By W. F. Butler. Illustrated.

VI. ON THE DESIGNING AND CONSTRUCTION OF STORAGE RESERVOIRS. By Arthur Jacob. Illustrated.

VII. SURCHARGED AND DIFFERENT FORMS OF RETAINING WALLS. By James S. Tate, C. E.

VIII. A TREATISE ON THE COMPOUND ENGINE. By John Turnbull. Illustrated.

IX. FUEL. By C. William Siemens. To which is appended the value of ARTIFICIAL FUELS AS COMPARED WITH COAL. By John Wormald, C. E.

X. COMPOUND ENGINES. Translated from the French of A. Mallet. Illustrated.

XI. THEORY OF ARCHES. By Prof. W. Allan, of the Washington and Lee College. Illustrated.

D. VAN NOSTRAND'S PUBLICATIONS.

XII. A PRACTICAL THEORY OF VOUSSOIR ARCHES. By William Cain, C.E. Illustrated.

XIII. A PRACTICAL TREATISE ON THE GASES MET WITH IN COAL MINES. By the late J. J. Atkinson, Government Inspector of Mines for the County of Durham, England.

XIV. FRICTION OF AIR IN MINES. By J. J. Atkinson, author of "A Practical Treatise on the Gases met with in Coal Mines."

XV. SKEW ARCHES. By Prof. E. W. Hyde, C. E. Illustrated with numerous engravings, and three folded Plates.

XVI. A GRAPHIC METHOD FOR SOLVING CERTAIN ALGEBRAIC EQUATIONS. By Prof. George L. Vose. Illustrated.

XVII. WATER AND WATER SUPPLY. By Prof. W. H. Corfield, M. A., of the University College, London.

XVIII. SEWERAGE AND SEWAGE UTILIZATION. By Prof. W. H. Corfield, M. A., of the University College, London.

XIX. STRENGTH OF BEAMS UNDER TRANSVERSE LOADS. By Prof. W. Allan, author of "Theory of Arches." Illustrated.

XX. BRIDGE AND TUNNEL CENTRES. By John B. McMasters, C. E. Illustrated.

XXI. SAFETY VALVES. By Richard H. Buel, C. E. Illustrated.

XXII. HIGH MASONRY DAMS. By John B. McMasters, C. E. Illustrated.

XXIII. THE FATIGUE OF METALS, under Repeated Strains; with various Tables of Results of Experiments. From the German of Prof. Ludwig Spangenberg. With a Preface by S. H. Shreve, A. M. Illustrated.

XXIV. A PRACTICAL TREATISE ON THE TEETH OF WHEELS, with the theory of the use of Robinson's Odontograph. By S. W. Robinson, Prof. of Mechanical Engineering, Illinois Industrial University.

D. VAN NOSTRAND'S PUBLICATIONS.

XXV. THEORY AND CALCULATIONS OF CONTINUOUS BRIDGES. By Mansfield Merriman, C. E. Illustrated.

XXVI. PRACTICAL TREATISE ON THE PROPERTIES OF CONTINUOUS BRIDGES. By Charles Bender, C. E.

XXVII. ON BOILER INCRUSTATION AND CORROSION. By F. S. Rowan.

XXVIII. ON TRANSMISSION OF POWER BY WIRE ROPE. By Albert W. Stahl.

XXIX. INJECTORS; their Theory and Use. Translated from the French of M. Leon Pouchet.

XXX. TERRESTRIAL MAGNETISM AND THE MAGNETISM OF IRON VESSELS. By Prof. Fairman Rogers.

XXXI. THE SANITARY CONDITION OF DWELLING HOUSES IN TOWN AND COUNTRY. By George E. Waring, Jr., Consulting Engineer for Sanitary and Agricultural Works. Illustrated.

XXXII. CABLE MAKING FOR SUSPENSION BRIDGES, as exemplified in the Construction of the East River Bridge. By Wilhelm Hildenbrand, C. E.

XXXIII. MECHANICS OF VENTILATION. By George W. Rafter, Civil Engineer.

XXXIV. FOUNDATIONS. By Prof. Jules Gaudard, C. E. Translated from the French, by L. F. Vernon Harcourt, M. I. C. E.

XXXV. THE ANEROID BAROMETER, ITS CONSTRUCTION AND USE. Compiled by Professor George W. Plympton. Illustrated.

XXXVI. MATTER AND MOTION. By J. Clerk Maxwell, M. A.

XXXVII. GEOGRAPHICAL SURVEYING. Its Uses, Methods and Results. By Frank De Yeaux Carpenter, C. E.

XXXVIII. MAXIMUM STRESSES IN FRAMED BRIDGES. By Prof. Wm. Cain, A. M., C. E.

*_** *Other Works in preparation for this Series.*

THE UNIVERSITY SERIES.

No. 1.—ON THE PHYSICAL BASIS OF LIFE. By Prof. T. H. Huxley, LL. D., F. R. S. With an introduction by a Professor in Yale College. 12mo, 36 pp. Paper covers, 25 cents.

No. 2.—THE CORELATION OF VITAL AND PHYSICAL FORCES. By Prof. George F. Barker, M. D., of Yale College. 36 pp. Paper covers, 25 cents.

No. 3.—AS REGARDS PROTOPLASM, in relation to Prof. Huxley's Physical Basis of Life. By J. Hutchinson Stirling, F. R. C. S. 72 pp., 25 cents.

No. 4.—ON THE HYPOTHESIS OF EVOLUTION, Physical and Metaphysical. By Prof. Edward D. Cope. 12mo, 72 pp. Paper covers, 25 cents.

No. 5.—SCIENTIFIC ADDRESSES :—1. On the Methods and Tendencies of Physical Investigation. 2. On Haze and Dust. 3. On the Scientific Use of the Imagination. By Prof. John Tyndall, F. R. S. 12mo, 74 pp. Paper covers, 25 cents. Flex. cloth, 50 cents.

No. 6.—NATURAL SELECTION AS APPLIED TO MAN. By Alfred Russel Wallace. This pamphlet treats (1) of the Development of Human Races under the Law of Selection ; (2) the Limits of Natural Selection as applied to Man. 54 pp., 25 cents.

No. 7.—SPECTRUM ANALYSIS. Three Lectures by Profs. Roscoe, Huggins and Lockyer. Finely illustrated. 88 pp. Paper covers, 25 cents.

No. 8.—THE SUN. A sketch of the present state of scientific opinion as regards this body. By Prof. C. A. Young, Ph. D., of Dartmouth College. 58 pp. Paper covers, 25 cents.

No. 9.—THE EARTH A GREAT MAGNET. By A. M. Mayer, Ph. D., of Stephens' Institute. 72 pages. Paper covers, 25 cents. Flexible cloth, 50 cents.

No. 10.—MYSTERIES OF THE VOICE AND EAR. By Prof. O. N. Rood, Columbia College, New York. Beautifully illustrated. 38 pp. Paper covers, 25 cents.

VAN NOSTRAND'S
ENGINEERING MAGAZINE.
Large 8vo, Monthly
Terms, $5.00 per annum, in advance.
Single Copies, 50 *Cents*.

First Number was issued January 1, 1869.

VAN NOSTRAND'S MAGAZINE consists of Articles, Original and Selected, as also Matter condensed from all the Engineering Serial Publications of Europe and America.

TWENTY VOLUMES NOW COMPLETE.

NOTICE TO NEW SUBSCRIBERS.—Persons commencing their subscriptions with the Twenty-first Volume (July, 1879), and who are desirous of possessing the work from its commencement, will be supplied with Volumes I to XX inclusive, neatly bound in cloth, for $53.00. Half morocco, $80.00. Sent free by mail or express on receipt of price.

NOTICE TO CLUBS.—An extra copy will be supplied, gratis, to every Club of five subscribers, at $5.00 each, sent in one remittance.

This magazine is made up of copious reprints from the leading scientific periodicals of Europe, together with original articles. It is extremely well edited and cannot fail to prove a valuable adjunct in promoting the engineering skill of this country.—*New York World*.

No person intererested in any of the various branches of the engineering profession can afford to be without this magazine.—*Telegrapher*.

The most useful engineering periodical extant, at least for American readers.—*Chemical News*.

As an abstract and condensation of current engineering literature this magazine will be of great value, and as it is the first enterprise of the kind in this country, it ought to have the cordial support of the engineering profession and all interested in mechanical or scientific progress.—*Iron Age*.

www.ingramcontent.com/pod-product-compliance
Lightning Source LLC
Chambersburg PA
CBHW022111160426
43197CB00009B/980